Luminos is the Open Access monograph publishing program from UC Press. Luminos provides a framework for preserving and reinvigorating monograph publishing for the future and increases the reach and visibility of important scholarly work. Titles published in the UC Press Luminos model are published with the same high standards for selection, peer review, production, and marketing as those in our traditional program. www.luminosoa.org

Muddy Thinking
in the Mississippi River Delta

Muddy Thinking
in the Mississippi River Delta

A Call for Reclamation

———

Ned Randolph

UNIVERSITY OF CALIFORNIA PRESS

University of California Press
Oakland, California

Suggested citation: Randolph, N. *Muddy Thinking in the Mississippi River Delta: A Call for Reclamation*. Oakland: University of California Press, 2024. DOI: https://doi.org/10.1525/luminos.183

Cataloging-in-Publication Data is on file at the Library of Congress.

ISBN 978-0-520-39720-0 (pbk. : alk. paper)
ISBN 978-0-520-39721-7 (ebook)

33   32   31   30   29   28   27   26   25   24
10   9   8   7   6   5   4   3   2   1

To Jessica and our girls, Annie and Polly

CONTENTS

ILLUSTRATIONS

ACKNOWLEDGMENTS

Muddy Thinking has been the culmination of many years of graduate and post-graduate work that started in San Diego and ended in New Orleans—two very different kinds of places. Throughout, I have carried the steadfast support of my graduate adviser and mentor, Patrick Anderson, who embraced the potential for *mud* from the outset and simply refused to allow me to stray off into more conventional paths. His introduction to the editors at University of California Press, I believe, was critical for keeping the idea alive and real for me—which it remained through the stewardship of three project editors, who passed the baton flawlessly. My wife, Jessica Shank, has been steadfast in her unerring belief in my ability to complete a dissertation, and now a book. Her support has been crucial, even if at times a little baffling to me. But I accept it as one excepts a very generous gift— with humility. My mother, Sanna Thomas, and stepfather, John Thomas, have been dogged copy editors with precision I have appreciated throughout. They proofread more than their share of work, which, I'm sure, exceeded what they signed up for. A publication such as this has gone through many iterations and readers. I can't begin to name them all. The other members of my dissertation committee, Valerie Hartouni, Kelly Gates, Angela Booker, and Octavio Aburto, as well as the University of California, San Diego, professors and fellow graduate students from the Department of Communication helped shape my thinking through conversations, seminars, workshops, campus pubs, and general care. I'm so grateful for them. My friend and former officemate, Alex Dubee, was especially supportive and available for close readings and feedback during my tenure at UC San Diego. I'm likewise grateful to my colleagues in New Orleans who helped me grow the work into a book. My friend, David Terry, planted the idea of muddy thinking in my mind at some point during the formal book proposal, which seemed to click everything into place. I would also like to acknowledge the financial support from the Bywater Institute at Tulane's Studio in the Woods, where I spent a week completing the manuscript draft, the Center for the Gulf South at Tulane, and the Institute of Practical Ethics and Judith and Neil Morgan Endowed Fellowship, both at UC San Diego, as well as the generous support from my department to travel and conduct research. My time as a visiting assistant professor at Tulane University also helped introduce me to the intellectual and activist community in South Louisiana that tirelessly advocates for a clean and equitable place to live. They continue to be my inspiration, especially the many grassroots organizations formed by and in solidarity with residents directly affected by pollution and environmental desecration. Finally, I would like to acknowledge my deceased father, Ned Randolph Jr., who visits these pages and my thoughts. I attribute the best parts of me to his gentle guidance and friendship. I wish he were here to stash a copy of *Muddy Thinking* on his bookshelf.

John McPhee's *Control of Nature*, published in 1989, documented the struggle of corralling the Mississippi River in a work of prose that first set spark to my literary imagination. *Muddy Thinking* might not have happened otherwise.

The following acknowledgment comes from the Studio in the Woods on the Mississippi River outside of New Orleans, where I spent a week in retreat completing my manuscript.

The earth here is built up from the alluvial soil carried by the Mississippi River. It was home to early native peoples well before European settlement. Indigenous communities include the Washa and Chawasha, the Houma, Chitimacha, Biloxi, Choctaw, Bayagoula, Quinipissa, Atakapa-Ishak, Caddo, Tunica, Natchez, Tchoupitoulas, Tangipahoa, and others. The city we call New Orleans has been a site of exchange and commerce for more than 1,000 years and was known among other names as Bulvancha—"the place of other languages"—in Choctaw. The Mississippi River takes its name from the Anishinaabe, a people native to the headwaters of the river, whose word Misi-ziibi, means "huge river." The diversity and richness of the deltaic ecosystem drew indigenous peoples, who lived with the rhythms of the river, traversing Bulvancha's ridges and bayous to hunt and to trade with each other. The creativity and ingenuity of both the indigenous and enslaved African inhabitants of this area are embedded in every aspect of what we now refer to as "New Orleans" culture and landscape—including music, art, engineering, agriculture, floodways, languages, and the ways we are in community together.

Introduction

A Turn to Mud

My early memories are murky and, like sediment, can be reshaped and sometimes permanently obscured. Particularly my memories of mud—alluvial silt and clay delivered by water so omnipresent in a childhood of bayous and streams. I grew up in a town on the Red River, which gave my home parish (not county) the name Rapides. Before the Red River was irreparably tamed by the US Army Corps of Engineers' $2 billion lock and dams project in the 1980s, it flowed with force.[1] It was too dangerous for swimming. And sometimes, when the water level was low, you could see the remnant earthworks of "Bailey's Dam," constructed under Union Lt. Col. Joseph Bailey during the Red River Campaign. The dam lifted the water for Union gunships to pass downstream. My dad, who seemed to know everything, said the dam was built to slow the pursuit of Confederates. In that campaign, Alexandria's downtown riverfront was burned to the ground. But Tecumseh Sherman, known for his scorched march to Atlanta, reportedly sent orders to spare the Louisiana State Seminary of Learning and Military Academy (later renamed Louisiana State University [LSU]), where he had served as its first superintendent.[2] He also spared several rural plantation homes as a sentimental gesture to the owners he had befriended during his post.

According to Alexandria's town history, a wealthy Pennsylvania landowner named Alexander Fulton laid out the city in 1805 after receiving a land grant from Spain two decades earlier. He named it after his daughter.[3] A different Fulton two years later would pilot the first steamboat up the Hudson River in New York, crossing a metaphorical Rubicon that would eventually open the Mississippi River basin to commerce and much of its southern tributaries to plantation slavery and Indigenous displacement. But I want to clarify that this is not a story about the Mississippi and Red Rivers—or at least not only about them. That story is famously retold with each new release of Mark Twain's canon or travel article about New Orleans or even the scratchy recordings of Jellyroll Morton and Bessie Smith and the lamentations in Paul Robeson's "Old Man River" about the unrequited nature of it all. John McPhee added his own imprint on the unintended consequences of controlling nature. The common narrative about the Mississippi River arguably comes from a *bias* of water. This project asks: What would happen if we start from a slightly different perspective? Would destabilizing and disorienting the landscape as it is popularly conceived bring forth questions that are not being asked in the deluge of water?

The book will certainly discuss water and land and how places and the people who live there are shaped by efforts to control nature. It will also explore how things may have been otherwise. The heart of this story is really *mud*. We will roll up our sleeves and get dirty, in a good way. This story proposes the framework "Muddy Thinking" to recast and denaturalize some of the effects that modern engineering and thinking have imposed on rivers and lands that have brought humanity to the edge of planetary extinction. But it is not meant to outline a dystopian future that forecloses discussions about possible action and alternatives. The specter of extinction is not the end of the story but rather a part of its "ongoing," as Donna Haraway would say. Extinction is an extended plateau of events. It is a long and slow process that "unravels great tissues of ways of going on in the world" for many animals and people.[4] As we journey along our current spectrum of history, extinction challenges us to respond. And *how* we respond is the question of our time. The provocation of this work aligns with what Haraway identifies as an ethos (my word) of *compost* (her word). This work investigates the entangled histories of people, racial capitalism, and mud.

THE BIG MUDDY

In this book, organized around New Orleans and South Louisiana as a case study, I pose a deceptively simple question: How could this muddy place, whose land and people are uniquely vulnerable to sea level rise and environmental injustice, be one of the nation's most promiscuous producers and consumers of fossil fuels? What cultural work makes this painful paradox feel not only possible, but inevitable? To answer this question, I bring together conversations in environmental studies and humanities to understand global warming as a technical and cultural phenomenon.

Once described in a *New York Times* article as a "disaster laboratory," Louisiana offers a compelling template for the contradictions of modernity and extractive capitalism.[5] The state's eroding shores, pollution, and petro-capitalism are emblematic of the forces causing global climate change. A three-century project to drain and reshape the Mississippi River since the colonial founding of New Orleans has been driven by interests to enable waterborne commerce, "reclaim" riverine marshes for plantation agriculture, and supply petrochemical plants with abundant feedstock of oil and gas. The harm of these practices is measured in "football fields" of land loss as well as high morbidity rates for minority communities on the fence lines of petrochemical and industrial plants along the Mississippi River corridor. In numbingly familiar statistics to residents, the US Geologic Survey estimates the state loses 45 square miles of coast-line a year—the equivalent of a football field every one hundred minutes—faster than anywhere in the world. Louisiana has lost more than 2,000 square miles of coast-land since 1930. Names of drowned waterways and villages disappear from updated maps along with estuaries for fisheries, seafood, and international migratory flyways. "Coastal Louisiana experiences some of the highest subsidence rates worldwide, mak-ing the Mississippi River Delta one of the first areas to experience the effects of global sea-level rise."[6] Rising seas—and the intensification of more frequent hurricanes that roll up marshlands—is accelerating this retreat and leaving New Orleans increasingly vulnerable behind levees while working-class hamlets, Indigenous communities, and other coastal villages sink.[7] If there is a place that shouldn't need convincing that the status quo is unsustainable, it is here in Louisiana. And yet authorities are hellbent on doubling down on the same old thinking, to the detriment of the residents who suffer the brunt of these processes. The state supplies 90 percent of the nation's offshore oil and gas infrastructure, which also feeds a secondary market of petrochemical plants up and down the Mississippi River corridor, known by residents as "Cancer Alley" or "Death Alley." Erosive oil and gas canals channel seawater into brackish estuaries. Spills happen with regularity, many with little public awareness. Thousands of miles of pipelines running through the increasingly disappearing coastal marshes face exposure to severe storms, and emptied oil wells subside underwater. Over a million permitted oil wells have been drilled in the state of Louisiana. There are over 50,000 active oil and gas wells and another 22,000 to 28,000 that are idle and effectively abandoned. The state counts another 4,628 wells that are documented as "orphaned," meaning no owner could be identified or had a plan for plugging them. These dete-riorating wells leak oil, methane, and saltwater into the ground and air.[8]

Capping and cleaning up only the orphan wells would cost taxpayers an esti-mated $400 to $560 million.[9] Plugging the approximately 28,000 nonproducing wells would cost $3.5 billion in closure costs alone, according to the Environmental Defense Fund.[10] But abandoning infrastructure has long been a national pastime for the oil and gas industry. There are more than 81,000 officially designated orphan wells across the country. A 2021 report published by the US Government Account-ability Office found that oil and gas producers have been allowed to abandon 97 percent of offshore pipelines in the Gulf of Mexico without incurring any pen-alties.[11] The effects of this corrosive infrastructure on vegetation—seagrasses and other subtidal species along the coast—are cumulative and largely unmeasured.[12] Meanwhile, Louisiana has one of the most concentrated industrial clusters in the world policed by a perennially underfunded regulatory agency in the form of the Louisiana Department of Environmental Quality that relies on self-reporting by plants that literally mail reports to the office, which are simply scanned by employees.[13] Unending river dredging and levee building is required to maintain the state's five deepwater ports that reside in an uninterrupted chain along the Mississippi River's banks from Baton Rouge to the Gulf of Mexico. Levees seal off coastal marshes from seasonal avulsions of mud and sediment; and dredging ensures the material flows farther down the river toward the Gulf of Mexico. Invasive species from international cargo lay waste to land-building vegetation and roots. And the weight of the Mississippi riverbed itself presses down as seas rise.

While Louisiana offers a cautionary tale of the destructive and dehumanizing effects of modern industry on land and people, it provides an opportunity to interrogate the deep, commonsense structures of what I call Extractive Thinking. Ironically, these same extractive forces that have led to coastal dissolution have also embedded themselves in solutions for coastal restoration.

By the time Hurricanes Katrina and Rita destroyed 200 square miles of Louisiana marshlands in 2005, the state had already lost 1.2 million acres of wetlands in seventy-five years. After Katrina, the state legislature approved the $50 billion "Comprehensive Master Plan for a Sustainable Coast," a plan partially funded by oil and gas royalties on federal offshore leases and a legal settlement from the historic 2010 BP Deepwater Horizon blowout. When I think of the local and global disconnect between fossil fuel production and local sea level rise, the words of an emergency manager from coastal Lafourche Parish haunt me: "If we don't have an economy, then what is there to protect?"[14]

I wonder if the solution is so irreconcilable. The presumption that oil is the only lifeblood of the economy ignores its falling employment numbers as well as generations of communities who have carved out their own covenant with the land. There is a pernicious temptation to keep to the path of extraction, particularly for an area so degraded by an industry that has consumed not only the land but also hope for a different future. Let this then be an example of what awaits others as climate change continues to render this planet less hospitable. Let us not forget the promise the Earth once held, before it disappears as a reminder of what was possible. "There is a fine line between acknowledging the extent and seriousness of the troubles and succumbing to abstract futurism and its effects of sublime despair and its politics of sublime indifference," writes Haraway.[15] In other words, we need to tell the difficult stories while also imagining new ways of living.

HOT TIMES

Just in the years since Katrina, hurricanes here have grown fiercer and more frequent as the Gulf of Mexico warms and Louisiana coastal marshlands disappear. Hurricanes now produce their own fuel as they approach saturated marshland in a phenomenon called the brown ocean effect. As a storm surge pushes warm Gulf water over inundated marshlands, it creates its own energy source. Rather than a buffering obstacle that protects inland communities, the marshland becomes an accelerant that increases wind speeds significantly just before it reaches populated areas like New Orleans.[16] In 2021, Hurricane Ida, for example, remained a hurricane sixteen hours after landfall and left little time to evacuate.

"Hurricanes draw their energy from warm ocean waters. But when they make landfall over a wet, marshy, or saturated spot, they can still power themselves with evaporating moisture."[17]

As this book was being finalized, we learned that July 2023 was the hottest month on Earth since record keeping began in the nineteenth century. And the ocean is getting warmer by the year. Since scientists have been keeping climate data, the Earth's hottest years have all occurred since 2015. The year 2021 was the ocean's hottest for the third year in a row, and the Earth's average carbon dioxide output was the highest ever recorded.[18] This occurred during a relentless global pandemic. A "Code Red" report on humanity was issued by the usually staid and cautious United Nations Intergovernmental Panel on Climate Change that pointed to the inevitability of increasing temperatures at an accelerated pace.[19] If all emissions had halted in 2021, the planet would still be warming. "We are decades late" making necessary changes, Kristina Dahl of the Union of Concerned Scientists testified in January 2022.[20] Four months later, in May 2022, the National Oceanic and Atmospheric Administration (NOAA) reported that its monitoring station in Mauna Loa, Hawaii, had measured an average level of carbon dioxide in the air that had not been seen on this planet since the Pliocene era 4.5 million years ago, when sea levels were 16 to 82 feet higher and temperatures were seven degrees hotter. "South Florida, for example, was completely under water. These are conditions that human civilization has never known."[21] Heat-trapping carbon dioxide has risen 50 percent since the pre–Industrial Revolution year of 1750. By the time you read this, things will be worse.

The world puts about 10 billion metric tons of carbon in the air each year, but a dramatic spike has occurred in heat-trapping carbon emissions just since 1990. Greenhouse gas pollution caused by human activities trapped 49 percent more heat in the atmosphere in 2021 than in 1990, NOAA reports.[22]

Have we have passed the tipping point already? Carbon dioxide, for example, remains in the atmosphere for hundreds of years. But curbing other greenhouse pollutants like methane, which has a much shorter atmospheric presence, could show immediate benefits. Expanding natural sequestration sinks through reforestation would also remove legacy carbon. Yet we continue toward the point of no return. We need to start thinking about how to manage ourselves in the ruins socially, ethically, and emotionally. We must figure out how to live and love in the shadow of doom *and* how to intervene effectively.

There is a temptation to fall back to the spectacle of disaster with a cynical detachment, to deny ourselves the emotional connection to a reality too frightening to contemplate. The *New York Times* culture critic Amanda Hess lamented that the end of the world we are experiencing "does not resemble the ends of religious prophecies or disaster films." There are no dramatic finales. "Instead, we persist in an oxymoronic state, inhabiting an end that has already begun but may never actually end." Social media's "apocalyptic drumbeat" of hopelessness ironically becomes a narcotic for it: "*Just hit us with the comet already.*"[23]

Meanwhile, current proposals to reach net-zero carbon seem to require little structural change to our notions of productivity and growth. Instead, they hinge on burying the evidence. Witness the Democratic Louisiana governor, John Bel Edwards, admonishing efforts by the Biden administration to impose a moratorium on drilling in the Gulf of Mexico just weeks before the governor delivered a speech at the 2021 United Nations Climate Change Conference in Glasgow where he touted Louisiana's initiatives to capture industrial carbon with new industrial projects.[24] Between 2020 and 2023, there were fifteen announced low-carbon, carbon capture, and so-called blue hydrogen projects announced in Louisiana—all of which would add to net emissions.

Yet, to his credit, the governor in 2020 appointed a climate task force—alone among southern states—to commit to reaching net zero by the year 2050 in line with the 2015 Paris Climate Agreement. But many of the environmental advocates on the task force point out a discomforting truth about solutions proposed by industry task force representatives. The efficacy of so-called industrial carbon capture for removing CO_2 is minimal—despite the publicity by big oil companies. Fossil fuel producers claim they can recover carbon dioxide emissions from smokestacks and store it permanently underground (or use it to recover oil and make other products like petrochemicals), which would "recycle" carbon emissions. Despite the marketing by fossil fuel companies, carbon capture is a less efficient, more expensive, and more dangerous method for lowering greenhouse gas emissions than natural carbon sequestration projects like reforestation, which has a proven track record and is not predicated on industrial production.[25]

Governor Edwards followed his speech in Glasgow with the announcement of a massive carbon capture project by a Pennsylvania company to produce new sources of "blue ammonia," which would create a net increase of carbon by a project that was touted to reduce emissions. Blue ammonia may be even worse for climate change than simply burning natural gas because of the leakage of methane.[26] Other "environmentally conscious" industrial projects are popping up throughout the state, which would also have an impact on low-income, fence-line environmental justice communities in Cancer Alley. Louisiana is positioning itself to be the storage site for hazardous carbon waste for the nation, or what the activist Monique Harden calls a mecca for hazardous waste.[27] Notwithstanding the glaring weaknesses of relying on aging infrastructure—or building out new pipelines and storage facilities in environmentally and socially stressed areas—industrial carbon capture and other net-zero fantasies offer little substantive solutions beyond assuaging the growing anxiety that we are out of real solutions. This book seeks to illuminate this intransigent madness, as well as offer some suggestions for finding life in the ruins.

The critical humanities scholar Fredric Jameson famously wrote that it is easier to imagine the devastation of the Earth and nature than the end of late capitalism.[28] This book suggests that the limitation is as much cultural as technical. What faces New Orleans at the toe of Louisiana's boot, as well as all vulnerable frontiers, is character-

FIGURE 1. The ExxonMobil refinery in Baton Rouge on the Mississippi River, built in 1909, is one of the largest oil and gas processing facilities in the world. Louisiana has sixteen such refineries, none of them younger than fifty years old. Eight of them, incidentally, were found to be the worst water-polluting in the country—and five of them were among the top ten. A 2023 report by the nonprofit Environmental Integrity Project found that among the toxic polluters of nickel, selenium, nitrogen, ammonia, and "total dissolved liquids," Louisiana's aging refineries ranked at the top. ExxonMobil, for instance, ranked number 10 in the country for selenium discharges. There are two hundred industrial, petrochemical, and heavily polluting plants along the river corridor between Baton Rouge and New Orleans. Image available on Flickr through Creative Commons license by Jim Bowen.

ized by what Amitav Ghosh calls "a crisis of culture, and thus, of the imagination."[29] We are caught in a paradox. In Louisiana, state authorities tout the importance of Louisiana's "working coast" of extractive industries to justify investments to restore a coastline that can sustain them. Interventions for restoration are trapped on a path dependent on industrial consumption, which is emblematic of the very paradox and dialectic tension of modernity itself. Here is where this book intervenes. It argues that mud and *Muddy Thinking* might gum up the gears of modernity.

THE RUSE OF MODERNITY

The cultural sociologist Chandra Mukerji describes modernity as a culture of survival and reinvention. As both a historical marker and a mode of living, modernity is fueled by dreams of utopia that fail to account for their dystopian effects. Powered by fossil fuels that radically alter their environment even before they burn, modernity destroys as it offers hopes for progress. It is built on the ashes of its own creation—dialectically generative by its own destruction.[30] Imre Szeman and Dominic Boyer note in their salient introduction to an anthology titled *Energy Humanities* that our very subjectivities are tied to abundant, cheap fossil fuels: "We are citizens and subjects of fossil fuels through and through, whether we know it or not."[31] That includes our *imaginaries* that such abundant energy makes possible. "In no discussion of freedom in the period since the Enlightenment was there ever any awareness of the geological agency that human beings were acquiring at the same time as and through processes closely linked to their acquisition of freedom," writes Dipesh Chakrabarty.[32] Modernity is responsible for the creation of enlightened, self-actualizing subjects while rupturing the planetary system that supports them. It has colonized our imagination with ever greater reliance on technologies that veer further from acknowledging our humble dependence on the nonhuman world, as well as our communal need for one another. "Nonhuman forces and systems had no place in this calculus of liberty: indeed being independent of Nature was considered one of the defining characteristics of freedom itself."[33] Modernity, writes Michel Foucault, "is the will to 'heroize' the present." Such heroics are poeticized in anthems of achievement, Springsteen's "mansions of glory in suicide machines," an adolescent's rebellion, or the cosmopolitan fetish of passport entry stamps. Modernity is authored in skyward jet streams.

But there is another, even more nefarious tendency of this modernity in the way it categorizes and distills. Joseph Roach describes this as "a taxonomy of segregationist behavior." In his book *Cities of the Dead*, which investigates New Orleans and London, Roach argues that modernity—particularly the European Enlightenment—was based on separating spaces. Modern cemeteries and death were segregated into the unhygienic silence of the tomb. Cities of the dead were socially distinct from those who enjoyed the status of the living.[34] Likewise, civilization was strictly separated from wilderness, which, in the words of Thomas Hughes, was framed as a "second Eden, ready to be manifested through man's unique aspiration and inspiration."[35] This new modern man transcended the chains of natural cycles to write a history of progress, triumph, and freedom without acknowledging its cost. "Only those people who had thrown off the shackles of their environment were thought to be endowed with historical agency; they alone were believed to merit the attention of historians—other peoples might have had a past, but they were thought to lack history, which realizes itself through human agency," writes Ghosh.[36]

Modern humans rationalized their curation of wilderness to "complete what God had started." New Orleans started as a product of this imaginary. It was made possible by a philosophy of taming a wild landscape through rational governance of separating water from land. To be modern is to know the world through Extractive Thinking. As a theoretical framework, I argue, Extractive Thinking has led to the plundering of vital ecosystems and then mitigating the harms of such practices through newly imagined methods of extraction. It attempts to invent and consume its way out of its own crisis. As our polar ice caps melt, multinational oil companies leapfrog one another to drill for the reserves below. Here in Louisiana, the modern project is traced through a political ecology of clear-cut cypress timber and denuded coastal marshes in pursuit of the very energy to fuel this plunder. Extractive Thinking sought to erase the mud by creating distinctly separate realms of land and water.

MUDDY THINKING

This book is an attempt to disentangle our imaginaries from the modern impulses that are embedded into the logics of the fossil fuel industry. It provides a historical and cultural analysis of Extractive Thinking through case studies that reveal how this logic has stripped our landscapes and harmed its people. To do this, it uses a material analytic that shifts our commonsense understanding of progress. It starts with mud.

Did you feel it? The drab, unheralded, unwanted, unsung layer of unstable detritus that constitutes the very material foundation of New Orleans and South Louisiana. Always pushing against the modern project of New Orleans and Louisiana is the problem of mud.

Mud.

While the river is memorialized in the national and New Orleans imaginary, mud fails to mark any identity. Extractive Thinking signifies this material through attempts to erase or control it. It is viscerally known and discarded. Muddy sediment was listed in old boring logs as "swamp muck." It teamed with life and smelled of rotten eggs. Today it is classified as clay, muck, coarse sediment, and peat. It seems to hide within the discourse of the river as an underside to a binary or disruptive agent. Mud has not only been a discarded element, but it has indexed loathsome bodies and spaces. The marshy terrain of New Orleans, for example, was a contested site of discourses from the nineteenth-century Sanitary Movement and its drainage infrastructure.

Defined by its mixture with water, mud often sits outside of scientific and scholarly discourses of sediment materials. Mud is instead what soil, clay, and sediment are not. It is leftover detritus. It is resistant to easy categorization and standardization. It is also quite unpleasant. A more complex understanding and consideration of mud may allow messy edges to persist and even spread where possible. Through mud, this book critiques relations of social power. It unpacks the cultural and racial history of New Orleans and the Lower Mississippi Delta region. Through mud, it tells a story of both the natural environment and the social conditions and histories entangled within it.

Signified by the pithy, timeless counsel to let nature take its course, the framework of Muddy Thinking invites us to settle in and get comfortable in the muck. Muddy Thinking, metaphorically speaking, disrupts progress. It opens possibilities with inaction, with partial visibility, with acceptance of the unknown, unactualized, and unextracted. It is an antidote to consumption. It is to accept value in areas undefined and economies of exchange rather than growth. It is to drop the nature/human dichotomy and embrace what Raj Patel and Jason Moore call an ecology of interconnectivity and to align with Haraway's call for "compositionist" practices that can build new collectives.[37]

Muddy Thinking stubbornly resists the prevailing idea that the climate crisis will be solved with exciting new technologies, entrepreneurism, or green capitalism. It instead analyzes how such technocratic discourses naturalize and rationalize limitless growth. For instance, how did ineffectual (yet financially lucrative) schemes to achieve net zero become so dominant in our thinking while doing so little to conserve our dwindling resources? How do such programs instead function as commonsense strategies to rationalize uninterrupted consumption? Muddy Thinking interrogates the functional logic that we can consume our way out of the crisis—through strategies predicated on continuing, if not increasing, consumption. It skeptically investigates "greenwashing" campaigns that rehabilitate the reputations of the very companies and practices that fomented this crisis.

Corporate pledges to reach net zero by companies like Amazon, AT&T, and Walmart are undercut by revelations of campaign contributions to climate deniers.[38] Muddy Thinking argues for a full audit of energy consumption to unveil the true costs of energy—even green energy that promises sustainability with little behavioral change of consumers. Muddy Thinking troubles the liberal ideal. It also provides an accessible framework for taking on discussions about the Anthropocene as a planet-altering epoch initiated by mankind.

TROUBLE IN THE ANTHROPOCENE

As both a geologic category and a recognition of the social moment for which it is named, the Anthropocene is identified by the Great Acceleration of consumption after World War II. Its legacy is measured in the scars and isotopes of the geologic record of the planet. The term was coined in the early 1980s by the University of Michigan ecologist Eugene Stoermer. It picked up steam in 2000 when the Dutch Nobel Prize–winning atmospheric chemist Paul Crutzen joined Stoermer to propose that human activities were so devastating, impactful, and measurable as to merit a new geologic term for a new epoch following the Holocene.[39]

The twelve thousand years since the end of the last Pleistocene Ice Age have been credited with stable planetary weather that allowed the flourishing of large-scale agriculture and civilizations. But the new designation, the Anthropocene, has called for more debate about when and how emissions began accelerating. Some look back to the late seventeenth-century steam engine; others, to earlier logics of industrialization and bodily oppression perfected by plantation slavery. The designation Anthropocene as a geologic category may be legible to scientists, but critical scholars argue that it may further naturalize human activities without interrogating the uneven effects of global warming on marginalized populations. Some have countered with proposals for a Capitalocene or Plantationocene to acknowledge the role of globalized colonization and capitalism. I'll return to this debate in the conclusion, but for now, I'll use the Anthropocene as a placeholder because it is legible across disciplines, from scientific discourses to the humanities, to discuss the problem at scale.

Climate change and planetary extinction are simply so large, so complex and dynamic to be what Timothy Morton calls a "hyperobject"—meaning that it takes all faculties, cooperation, and vantage points to even frame and understand its complexity.[40] It appears differently to different perspectives and locations. It acts differently to different disciplines. It is subtle, dramatic, historic, imminent, large, and contradictory. The planet is warming, yet winter storms are more dramatic and abundant. Drought is chronic in large swaths of the globe as flooding simultaneously threatens. The Mississippi River in 2019 reached flood levels from record rainfall in the Midwest, prompting the US Army Corps of Engineers to open the Bonnet Carré Spillway in Louisiana for a record seventy-six days. Three years later, drought in the Midwest lowered water levels below partially buried shipwrecks and prompted the government in lower Plaquemines Parish, which relies on treated water from the river, to deliver bottled water for weeks to residents. The inability to reduce these contradictions to sound bites hides its reality. Reductionist thought is its weapon. As an unofficial moniker, the Anthropocene has currency among many different practitioners. It is also open to challenges as we discuss interventions and the stakes of inaction. The raw spectacle of terror that global warming incites can also lead to inaction. So balancing between poles of denial and self-defeating catastrophe is important but not easy. In fact, it is probably the more difficult approach.

STAYING WITH THE TROUBLE

I borrow the tagline, "Staying with the Trouble," from Donna Haraway to underscore the intellectual challenge of addressing and disrupting the power of Extractive Thinking. One of my biggest concerns in researching this project was a lack of legible solutions for transcending the paradigm. The dilemma continually eluded my conventional grasp, which makes sense. The exceptional challenge of climate change reflexively raises the question of how we can persevere without some technical breakthrough. There are too many of us and too few resources. How can modern people so accustomed to comfort and individuality puzzle a way out of this dilemma? What I continually, frustratingly, came up against is the limitation of the paradigm itself. There is no modern solution, at least not from this conventional vantage point. We must change much more than our automobiles and gas stovetops. If we expect battery-powered motors or wind and solar energy to save us, we remain on the same dreamlike trajectory. Our assumptions about unlimited energy must change. We must instead labor to find understanding in what Anna Tsing calls "life in the ruins."[41] Even as the ground beneath us trembles and shifts, we remain on the ground. Staying with the trouble requires learning to live in the present, no matter what it happens to resemble: not a "vanishing pivot between awful or Edenic pasts and apocalyptic or salvific futures." We must face our own continuing. Life in the ruins will also be shared with "mortal critters entwined in myriad unfinished configurations of places, times, matters, meanings."[42] This requires telling some pretty tough stories about the present.[43] We must resist looking away and dissociating ourselves from what is happening and will continue to happen. We cannot disconnect from an uncomfortable present and uncertain future. We cannot take comfort in the fact that in the future we'll be dead anyway, as former President Trump mused.[44] Rachel Carson sixty years ago wrote that we have an obligation to endure, which should also be understood as an obligation we have to not only ourselves, but others.[45]

If we can't invent our way out of the paradox of annihilation and survival, then perhaps we must let this modernist quandary die. We may look to other epistemologies and practices not predicated on extraction such as the insights of Indigenous peoples exercising care-based stewardship with a deep connection to place. The anthropologist Kristina Lyons studied Amazonian farmers who were living and working on land that was written off as fallow by the Colombian government. She wondered how such communities could sustain themselves and thrive in the midst of the threat of annihilation and war. What she found was that by not participating in the "high-modernist extractive policy of narco-eradication" or mining, they had instead carved out a transformative space within the dense entanglements of decomposing leaves and rootlets and the insects, small animals, and birds cloaked by selva canopy. "It was a tenacious vitality of life . . . pulsating away."[46] Modes of eating, seeing, cultivating, and decomposing allowed these ecologies to endure. "What I learned," Lyons writes, "was that rather than rely on productivity—one of the central elements of modern capitalist growth—the regenerative potential of these ecologies relies on organic decay, impermanence, decomposition, and even a robust fragility that complicates modernist bifurcations of living and dying."[47] In other words, life persists in decay. Life beyond modernity is possible.

On this trajectory of modern "failure," we will need a methodology that reframes what we think we know about the world. "Farms are never only farms when they are also always regional watersheds, foothills, forests, biological corridors, and floodplains."[48] We will need to interrogate how knowledge production about the environment—scientific research or even environmental journalism—constructs a particular object for human-centered utility. The *environment* implies fragility and limitations. How has our rhetorical frame affected how we experience and represent what we consider the nonhuman world? How has this logic brought *mud* itself into being?

In Louisiana, there are six major categories of land that are defined by how waterlogged they are. "In many cases the distinction is arbitrary as many areas represent transitions between the two."[49] Cypress swamps and marshes register the transition from freshwater habitats in the upper delta plain to brackish and saline habitats in the lower delta plain.[50] Areas that are less inundated become forested. Closer to the sea, much of the marsh is unwalkable flottant.[51] In essence, the taxonomy of southern Louisiana is a classification of mud, which is somewhere between land and water—a liminal state that resists stable classification. It is context-dependent.[52] "Biologists and ecologists have found that wetlands are difficult to define—they have identified thirteen types in all, and their boundaries are hard to define. They may be permanently inundated, seasonally inundated, intermittently inundated, or seasonally waterlogged."[53] Wetlands are so named because water saturation is the dominant factor determining the nature of soil development and the types of plant and animal communities living there, according to the US Department of Interior Fish and Wildlife Service.[54] In other words, wetlands are classified not only by what they *are* but also by what they *do*—which opens interesting questions of *ontology*. How we think about mud and water has a lot to do with how they are used and by whom they are framed.

Historically and even today, when political and business interests have discussed the Mississippi River, they have conjured up a body of water moving over land, which is not really what an alluvial river is. Alluvial rivers are silty. Riverbanks and riverbeds erode and move. Alluvial rivers bend, loop, and crevasse in unexpected directions based on paths of least resistance. When a riverbank floods, as the Mississippi River's often did, the river's muddy flow spilled into other geographic, social, and political arenas. Modern engineering and political impulses wanted to corral and stabilize the river. Levees were raised higher, and the river's confines were narrowed. These nineteenth- and twentieth-century efforts to discipline the river created new problems.

As engineers leveed, narrowed, and shortened the river, they turned it into a more efficient waterway: a self-scouring engine that became cataclysmic when levees failed, which they inevitably would do. Interventions in the Mississippi River have led to the largest "natural" disasters in American history, recounted in American literature, oral histories, news reports, geologic surveys, spirituals, blues recordings, ballads, journals, jail logs, plays, and civil rights complaints. This legacy also disrupted the ecological processes of the Louisiana delta formation that had taken place over several millennia. By the 1930s, researchers understood that the Mississippi's River's historical, geomorphic meandering had deposited thousands of layers of organic soil that nurtured a hardwood bottomland forest and built an alluvial delta. But as the decades in the twentieth century progressed, they began to suspect that the Herculean effort by the Army Corps of Engineers to dredge and levee the river to protect communities from flooding was choking off the Louisiana marshlands from their progenitor. As sediment and mud were jettisoned into the Gulf of Mexico, the adjacent, bypassed marshes were left vulnerable to other human-induced stressors, particularly intensive oil and gas drilling that left behind thousands of miles of canals and pipelines.

The cumulative effect had been obvious to locals for years. The wetlands and barrier islands were converting to open water. No one seemed to quantify this historical dynamism until the 1970s. Despite a landmark study in 1981 that linked coastal erosion to river control, the Army Corps of Engineers in 1994 was still officially doubting the link of wetland erosion to river controls and instead attributing it to natural seismic movement from submarine salt domes, geologic faults, and oil and gas canal spoil banks. The Army Corps of Engineers simply refused to consider its own work on the river as a cause of coastal erosion. At the other end of denial, the powerful energy lobby refused to accept its own causal role from cutting canals, drilling oil wells, and leaving behind toxic wastewater brines. As various actors pointed fingers at the other's culpability, the swamps and wetlands that for centuries had buffered communities from storms and sustained a rich ecology of seafood, flora, and migratory flyways continued to disappear.

KATRINA'S GHOSTS

Few contemporary narratives of New Orleans escape the thematic vortex of Hurricane Katrina. So it is with this account. Just about everyone has their own Katrina story. And its meaning changes over time. By the time the Category 3 hurricane churned into New Orleans, the swamps and marshes that protected the city from major storms off the Gulf of Mexico were long denuded and the concrete seawalls on the city's edge and interior drainage canals were neglected. The storm easily penetrated this weakened, and neglected, system, pushing water up dredged canals into an urban bowl that had for all intents and purposes constituted a three-hundred-year project of a modern imaginary known as New Orleans.

The long arc of survival for this city in a swamp had required an ongoing regime of cultural, political, and economic practices to stabilize the Mississippi River, drain swamplands, and build a fortification of levees in response to one crisis after another. In fact, such crises and responses to them organized much of the city's political economy and culture. Various efforts to come together to address this multicausal phenomenon fell flat or failed to properly scale to match the challenge. It wasn't until Katrina that the political stalemate was broken. State authorities were able to cobble together their case for recovery by leveraging the importance of the region's assets: a deepwater port, a seafood industry, and a fossil fuel industry, which includes a corridor of petrochemical plants and oil and gas infrastructure between New Orleans and Baton Rouge that account for more than a quarter of the nation's energy supply.[55] State officials would like us to remember all of these assets, which they continually use to rationalize the state's strategic importance in order to justify the vast federal resources needed to fund their master plan for coastal sustainability. "Technology is seldom an unmixed blessing. The levees that shield New Orleans also intensify the process that are consigning it to the Gulf,"[56] writes Todd Shallat. Therein lies the tragedy of Extractive Thinking.

It is a circular *discourse* that has become a natural commonsensical way of viewing the world. It is reinforced by technical interventions to manage the forces of nature, because these interventions generate—or regenerate—a particular kind of governance that continues to reproduce conditions for its necessity. Efforts to secure the people, economy, and culture paradoxically increase the existential threats against them due to the destruction of the land itself. By erasing and refashioning the mud for their own discrete purposes, to build levees or dredge the river and canals, administrators inch closer to the imminent demise of the entire project.

As a discourse, Extractive Thinking frames mud as a fungible object in a particular kind of water story that is narrated in reports of sedimentologists and geologists. It circulates among state restoration boosters who fetishize the river and erect a multimillion-dollar "water campus" to revitalize a downtrodden downtown area of Baton Rouge. Analyzing this discourse requires highlighting what is left out of many discussions about the Mississippi River: the vast sediment of silts, clays, and mud that is carried through the continental body, escaping through various entrapments designed to keep the water flowing for shipping, and targeted mainly to protect oil and gas infrastructure.

ENVIRONMENTAL FATALISM

As I undertook and continued to wrestle with this complex project, I was left wondering what effects controlling the Mississippi River had on those in the path of potential destruction. The natural order had long given way to engineering. How did living in the specter of disaster, both economic and existential, affect the worldview of those who depend on the state to hold the river in its course? How were they conditioned to view the arbitrariness of nature? At the other end of the spectrum lay the extreme pragmatic, if not cynical, account of a petroleum-dependent economy that appears to any outsider to be destroying the very land on which its participants lived. How was it possible that the biggest critics of environmental regulation during the British Petroleum oil spill in 2010 were those people in the very path of the spilling crude? Within days of capping the BP spill, thousands of residents gathered in the Lafayette Cajun Dome with T-shirts emblazoned with "Drill Baby Drill" to protest President Barack Obama's temporary moratorium on Gulf drilling to assess safety protocols.[57] From these two poles, I began to search for a common link. On the one hand, there was a population dependent on the state's tenuous hold of the status quo; on the other, there were those whose very way of life was organized around degrading their surroundings. Were they connected? If so, how? They seemed to be brought together in the shadow of environmental manipulation that in some way demystified and commodified the landscape. Did this produce a crude and pragmatic calculation of environmental fatalism? It seemed related and tied together, but what was the connecting braid? How did the landscape become such a basic, if fragile, utility whose purpose was only to provide resources?

What I came to suspect is that the long history of intervention in the river had been so "naturalized" that the possibility of the river resuming its prehistoric behavior of meandering came to be seen as unnatural. The late US senator from Louisiana, J. Bennett Johnston called it "unthinkable."[58] In this perverse perspective—something completely paradoxical in fact—the artificial becomes natural and the natural becomes unthinkable. And that is where we have found ourselves—to the point where classical economics and modern thought have failed to forge a solution. We can no longer think with the modern tools we have. In fact, our tools make less sense by the day. They are rendering our logics unworkable. And we can no longer think with the myriad other diverse agents in our midst.[59]

Our challenge will be learning how to exist with other beings also struggling to survive. We can focus neither purely on the economics nor purely on the *natural* ecology of the landscape. Instead, we need to learn think *with* one another. How will people subsist in degraded areas? How will people earn money to live in areas after resettling away from the coast? The cynical—and easy—answer is that the oil and gas industry will suck up the last viable drop while avoiding intensive safety upgrades and move on. By then, most of the other sustainable jobs as well as schools and community services will be long gone. Residents will have relocated because they couldn't afford higher insurance or were unable to finance a mortgage in disaster areas. Or perhaps they couldn't navigate a submerged coastal road to get to work. These things are happening now. Louisiana, for example, represents 10 percent of all US flood claims. The home owner's insurance industry is on the cusp of collapse in the state.[60] Legislators huddled in a special session called in February 2023 to try to lure insurance carriers back to the state after Hurricanes Delta, Laura, Zeta, and Ida generated 800,000 claims of $22 billion in damages between 2020 and 2021. The state-run insurer of last resort, Louisiana Citizens Property Insurance Corporation, which has become the only option for many residents, was set to boost rates by 63 percent in 2023 to remain solvent.[61] (My own home is on its third policy in as many years after each carrier has dropped our coverage to "reduce risk exposure.") Repeated disasters are dramatically changing the remaining riverine forest ecology faster than it can recover.[62] Meanwhile, the Corps of Engineers, at the behest of Louisiana economic officials, continues to dredge a 50-foot draft channel in the river that extends like a lone vein into sea.

But this is not a provocation to turn away. Can we find a way for people to maintain their ancestral homes in precarious regions without continuing the very extractive practices that are destroying them? Can we avoid environmental fatalism and imagine a landscape rich with untapped possibility? Can we move beyond a crude cost-benefit analysis for intervention that operates on reductionist logics, blind to externalities of environmental damage that both add costs and reduce benefits?

To explore these questions, this work brings together multiple disciplines to tell a particular story of this place. It is neither an exhaustive nor an authoritative account. It speaks with a diversity of contemporary and historical sources, both primary and secondary. It works within the archive while attempting to address the politics of knowledge production of who gets to speak and how and which spaces are silenced. The book's sometimes contradictory viewpoints, I believe, register the instability that is inherent in staking claims, both territorial and epistemological, on this landscape. The very boundary of Louisiana, where swamps and bayous give way to coastal marshes and rivers flow into oceans, is somewhere between land and water that has been debated and adjudicated since colonial settlement and US statehood. Its tidelands and waters are themselves in constant flux. Today the most dramatic changes come in the form of reclamation by the sea—set in motion by a combination of natural alluvial physics, climate change, and capital extraction. Put simply, it is hard to find firm footing here. Intellectually, it changes. I'm from here, but I don't always feel like I belong here. As I try to frame the current imaginary of this place by opposing interests, I have done my best to be fair. And I acknowledge that any inaccuracies or reproduction of accounts that have since been challenged or changed are my own responsibility.

While specific to its own eccentric locality, it exists as part of a global ecology. Ongoing climate change comprises many places of change, each unique to its own history. Writing from interviews, archival documents, and observations from field sites and workshops, I analyze how the river's delta, its mud, and its people have coevolved. I also don't suggest a fix for this dilemma. In fact, I believe such fixes are part of the modern reproduction of Extractive Thinking. A modern answer continues to commodify the coast for extraction—as has been done since the arrival of settler colonialists—to the detriment of marginalized and Indigenous peoples. This extractive logic has led to a federalized response to controlling the river and protecting New Orleans. It led to a new $14.5 billion levee wall around New Orleans, whose levees immediately began to subside and its pumps to corrode, and a partially funded $50 billion master plan.[63] It has tied Louisiana's future interventions to the oil and gas industry and industrial shipping that have led us here. If centuries of discursive practices are embedded in the vast artillery of dams, levees, jetties, and spillways that produced a deep and swiftly moving Mississippi River, what would it mean to reread the history of New Orleans and the Mississippi River through the optics of mud? What kinds of naturalized discourses could a genealogy of mud dislodge? This book gestures toward questions that examine the rationales, ideologies, and culture located on this spectrum of tension. I examine how ongoing interventions bolster the status quo in the name of security—both material and economic—and prompt additional measures of security. One of the things the project will be considering is how the logic of extraction became naturalized in economic, political, and scientific thought over the past three centuries. It is through this naturalization process where both extraction and restoration have become so entangled that they are part of the same conundrum.

Muddy Thinking may not be the answer to the problem of modernity. But we might apply here what Anna Tsing so aptly observes about Matsutaki mushroom picking: "We are stuck with the problem of living despite economic and ecological ruination." It is time to pay attention to the mud, just as it is with mushroom picking. "Not that this will save us—but it might open our imaginations."[64]

Interlude

Vignette

It is maybe late summer or fall. It's dusk, I'm sure. The day is growing late and quiet, even as I hear my mother call across the water. I take another step out, and my foot sinks into red clay. The water seems to fold over itself in sheets. Another step, and my foot sinks farther into the mud. It reaches up close to my knee. I was following my dad, I think. I wanted to catch him, maybe to be carried. Water begins to swirl around my legs. "Stay there!" I can hear her calling, now from behind me. She is lying down on the bank, her back has seized up. We had fallen off the mare earlier. We were bucked suddenly while walking her up the levee. We fell and rolled. Something in the grass had spooked the horse. Mom went down first on her back into the tall grass. I fell on her. She shielded me. I was okay. But she never fully recovered, chasing chiropractors and holistic healers the rest of her life.

She is calling louder now, pleading for me to stop walking. The water is high, and my feet sink deeply into the mud as the currents of the Red River swirl around my knees. I lift my foot painstakingly out of muck. My foot comes up with a sock red stained with clay. The river mud has swallowed my shoe. As I start to cry, I feel myself swooped out of the muddy riverside by my dad and carried back to the bank. "I thought I'd lost you," she would say over and over. "I thought I'd lost you." Another murky memory.

Soon I am in the tub at my grandmother's house a mile inland from the riverbank. The warm bathwater cleans off the red mud from my socks. My clothes are being peeled off in the tub now ringed with red and brown sediment. "I was so scared," I can hear Mom say. The water is warm and calming. By the time I write this years later, the Red River will have been wrestled away from nature, its color taken, and its fierceness dormant. And later after reading this passage, my mother recalls, "One more step, and you would have been swept away. Mud saved you."

A Mudscape in Motion

To illustrate how we might use the framework of Muddy Thinking, I will start with the unassuming perspective of mud and its powerful delivery system: the Mississippi River. I will reframe the river, not as a waterway bordered on either side by land, but as what the Army Corps of Engineers historian Todd Shallat calls a *mudscape* that constructed the alluvial delta on which South Louisiana sits. Many thinkers and spiritualists, philosophers, and scientists have noted a peculiar commonality among rivers. Paraphrasing Heraclitus, no one ever steps in the same river twice, for it is not the same river, and they are not the same person. A river is a locus of force and canvas of imagination. A river is always in motion. It is not so much an object but a procession where things happen. In this spirit, I open the chapter through the perspective of the river's most basic assemblage: the mud that constitutes it.

Imagine a collection of fragments: granite dust, animal waste, decomposed bone, flakes from an ancient pot shard collected into a fist-sized lump that slides down into a rain puddle. Let us give this lump the name "mud." Imagine the mud stuck to the leg of a bear that has wandered through. Dried now but with its structure intact, mud's interloper finds her way to a river in search of food. Now freed and reanimated by water, our mud begins to move—now pushed, now pulled—into an alluvial flow that empties as it travels through and around a cluster of stones and trees into a tributary or brook whose immediate aim is to join with other sediments and flows. By now, our lump of mud has grown, collecting unto itself a series of other castoffs: the remnants of rotted leaves, larvae, and feces.

Let us remember that the lump is alive, home now to a gathering of microbial agents whose work to digest its component parts continues with its interaction with hydrogen, carbon, and oxygen. This living lump now finds itself in a tidal pool: lurching, joining, and separating into a viscous torrent of carbon-based fragments of limbs, bone, food, waste, and affluent—a changing manifest of geography and history intermixed with what we have dreamed as the natural world. They are suspended in animation together in confluent currents, pitting and prying against stratified forces within the water column that swirl and eddy against themselves—always in the inevitable urge toward the sea through the forceful momentum of movement that began when a tectonic collision buckled the continent at its center and lifted the Appalachian Mountains 300 million years ago. Into the resulting crease, a great basin of tributaries flowed, taking the continental face with it.[1]

Epochs of glacial movement and withdrawal followed by regular seasonal snowmelts have poured waters and sediments into this cleft from as far as the Rockies to the west and the Alleghenies to the east. In these upper portions of the Mississippi River's tributary basin, the current moves swiftly through old glacial groves toward and into the crevice of the watershed. In the less rocky lower valley, the Mississippi broadened and slowed because it was unconstrained by hard ridges. Before the arrival of Europeans and their levees, the Mighty Mississippi is said to have resembled an inland lake—a "shallow and wide glade of free-flowing tributaries from the western and eastern corners of the northern continent"—that seasonally flooded for miles and nourished an expansive milieu of hardwoods: oak, ash, elm, willow, cottonwood, tupelo, cypress, and sweet gum; habitats of fish, shellfish, reptiles, amphibians, panthers, wolves, raccoons, otters, muskrats, and opossums; and great flocks of birds.[2] From the great confluence of the Missouri and Ohio Rivers, the Mississippi drained into a flat delta south of Cairo, Illinois, and meandered "like a pianist playing with one hand—frequently and radically changing course."[3] Like those cultural iconoclasts who created the rhythmic register of the Delta blues, the river's abundant payload overflowed constraining boundaries of the day—scouring and consuming its banks in unexpected shifts. "Each time ice accumulated on the continents and sea level fell everywhere, the Mississippi River cut a broad trench along its lower course. With each succeeding interval of glacial decay, sea level rose and the Mississippi partially filled the extensive valley, thus gradually creating white deposits of river borne sediments."[4]

The mud directed the stream where to go—surging over to the left, slowing against a ridge of its own making—from which to push off into new directions. "If the flood waters are left free to act as they will, no one can predict from year to year where the next year's low water channel will be found, nor what will be the least depth in the bars," wrote William M. Black, chief of the US Army Corps of Engineers in 1927. "Under such conditions, a river port of one year may find itself inland the next, and river carriers may have to be tied up indefinitely at wharves."[5] As the channel meandered, it dropped sediment that added to the resistance of the land by adding to the land itself. The last 450 miles of the river's flow lies below sea level, which means that river bottom currents have no reason to flow at all. "But the water above it does. This creates a tumbling effect as water spills over itself, like an enormous ever-breaking internal wave."[6] It attacked its banks like a buzz saw until it forced passage, resulting in a torrent that would become a cascade.

A century ago, the Lower Mississippi River carried 400 million metric tons of sediment to the lowlands and Gulf Coast every year—enough to cover the entire state of Louisiana in almost an inch of mud.[7] Shallat describes the Mississippi as "curling and coiling like a snake in a sandbox," bleeding soil from thirty-one states and weighing down on "butter-soft low-lands."[8] The amount of mud carried in the Mississippi has historically ebbed and flowed by the runoff that drains from the continental basin's rivers and streams.[9] Today an estimated 600,000 cubic feet of water and sediment flows down the Mississippi every second—equal to the 53 million residents in ten states along the river's watershed from Minnesota to Louisiana flushing their toilets twice a minute every day.[10] In the mid-nineteenth century, the Mississippi was called "the great sewer" by Mark Twain's mentor, Captain Marryat. "This mud, solidified, would make a mass a mile square and 241 feet high."[11]

The land south of Baton Rouge consists of alluvial delta that began forming over eight thousand years ago as the great sediment flow from melting glacial ice pushed south beyond the Pleistocene Ridge onto a muddy shelf of its own making, unevenly dropping and pushing mud, clay, sand, and silt into the body of water named by Spanish explorers Seno Mexicano.

SOCIO-NATURAL EVOLUTION

When the conquistador, Hernando de Soto, set out from Tampa Bay in 1539, he reached the eastern Mississippi shores in spring 1541. His expedition's quest for gold was slowed only by the river's intransigent muddy edges. Garcilaso de la Vega's direct account described a seasonal flood of the Mississippi. It was severe and prolonged—beginning March 10, 1543, cresting forty days later. The expedition of more than six hundred soldiers, retainers, captives, mules, horses, hogs, and hounds trudged through waist-deep marsh.[12] They waded through water, and when the river finally receded that summer and fall, they marched through mud. The next spring, the river returned, muddying the valley again. For de Soto's men, the "pathless forests" were sometimes too muddy and inundated even on horseback. But they were too shallow for boats.[13] The men were beset by a problem of muddy marsh that impeded their progress yet provided ample supplies of fish and nourishment for the encampments. Fish were so plentiful they were killed with clubs.

They also encountered, sometimes resistant, complex societies. Over the three-year expedition, the conquistadors exploited, killed, and made alliances with many tribes. But it did not turn out well for de Soto. Just after reaching the eastern shores of the Mississippi, the expedition was met by Aquixo, ruler of the province of Quizquiz, a linguistic Tunica community on the western shore of what is now the Arkansas River. The chief arrived with a fleet of two hundred dugout canoes outfitted with banners, shields, and warriors wearing colorful feathers. The party paddled across the river with Aquixo seated beneath a canopy over the back of a large canoe; Aquixo presented de Soto with three boatloads of fish and plum loaves.[14] According to some accounts, de Soto's men launched arrows against Aquixo's men out of fear. Other accounts claim that de Soto was initially fired upon.[15] Both versions foreshadow an uneasy and violent relationship.

The Tunica people claimed that Quizquiz was subject to an even more powerful leader, named Pacaha, who lived farther north. Five weeks later, the Spanish expedition successfully crossed the river on four large rafts they built from logs near present-day Memphis. "They entered the province of Casqui after two days of very difficult travel through swampy lands."[16] Pacaha several times sent presents of skins, shawls, and fish. The catfish alone weighed up to one hundred pounds. There were buffalo fish, paddle fish, large-mouth bass, bluegill sunfish, and freshwater drum "the size of hogs."[17] In March 1542, after a severe winter, de Soto moved down the Ouachita River from Arkansas into current Louisiana. He likely went southward to a large prosperous agricultural settlement named Anilco, at present-day Jonesville, Louisiana, and turned eastward to Guachoya on the Mississippi near Ferriday, Louisiana.

At some point, de Soto died, leaving the expedition in the hands of Luis de Moscoso, who led the "weary Spaniards" overland, where they encountered a Caddoan tribe. They then marched southwestward, crossing paths with more linguistic Caddoan tribes, and eventually reached the Mississippi River, covering a total of 200 miles through wetlands.[18] "In July 1543, the surviving Spaniards embarked on small boats they had built and set off downriver on the long trip to Mexico."[19] For several days they were pursued and attacked by tribes that were probably Natchez. Nearing the mouth of the river, they were confronted by others, likely Chawasha, Quinapisa, or Washa, because of their use of atlatls, ancient devices developed in Mexico to throw spears. The Spaniards encamped on Timbalier Islands and, by September, finally reached Mexico.[20]

The expedition's account stood for one hundred fifty years as the only written record of Western contact with Indigenous Amerindians. But by the time Great Britain established Charleston in 1670 and the French began their exploration of the Lower Mississippi Valley in 1673, the complex societies that the de Soto expedition encountered had vanished.[21] Our direct knowledge of the expedition comes from three narratives by the expedition's surviving Spaniards and later archaeological and environmental studies on the river's former routes that attempt to clarify interactions and subsequent devastation of the Indigenous chiefdoms. The thriving population observed by the Spaniards had shrunk by 80 percent. One theory is that they may have already been on the decline, hastened by the diseases introduced by the Europeans such as influenza, whooping cough, measles, cholera, smallpox, and the common cold. The Amerindians may have reorganized into more fragmented groups.[22] Research on trees has uncovered evidence of a drought at the time, as well as a possible "little ice age" of colder than average temperatures.[23] The towns on the west bank of the Mississippi were largely abandoned by 1700. Periodic increases in flooding, coupled with the desire to avoid older living sites that had been decimated by European diseases, may have pushed seventeenth-century Indigenous people out of the western flanks of the river onto the higher eastern bluffs in present-day Mississippi.[24]

The river itself, particularly in the lower delta, was in dynamic action, creating new land and shifting directions altogether—so much that René-Robert Cavelier, Sieur de La Salle, who "claimed" the river in 1682 for France when he sailed out of present-day Lake Pontchartrain, failed to locate the mouth of the river four years later.[25] He was following the cartographic writings of the de Soto expedition, which actually may have descended the Atchafalaya River, not the Mississippi.[26] Spanish and French chroniclers used entirely different names to identify the Indigenous Amerindians they encountered, which adds to the uncertainty.[27] La Salle was subsequently murdered in a mutiny after landing five hundred miles off course near present-day Galveston, Texas. Whether the mouth of the Mississippi shifted or was misdocumented, its change in appearance one hundred sixty years later is plausible given that the river and mud were forging new meander paths to the sea. The current path of the river downriver from New Orleans is by some estimates between 350 and 600 years old.[28]

WORKS PROGRESS ADMINISTRATION ARCHAEOLOGY

Digs funded by the Works Progress Administration (WPA) in the 1930s—which initiated modern archaeology in Louisiana—have identified sophisticated Indigenous settlements along the Mississippi River's former path lobes and meander routes.[29] The earliest settlements along the Mississippi River are some of the oldest in North America, dating to 4,000 BCE. These Early Archaic sites exist near springs, around swampy lakes, and alongside hills and rivers. "Such locales provided deer and abundant small game, fresh water, vast quantities of beech, hickory, oak and Chinquapin mast, stone for tools, and hardwood for implements and fuels." There were also salt workshops, produced by boiling spring waters around large salt domes in northwestern Louisiana near Shreveport and in South Louisiana among the "five islands" perched atop great salt domes: Avery, Weeks, Cote Blanche, Belle Isle, and Jefferson.[30]

During the Archaic period, projectile points evolved from large, fluted tips to shorter, wide, notched heads. Cutting edges were erratic, and notches gradually evolved into large barbs. "These are the familiar points that fill bags and boxes in countless private collections all over Louisiana. Few of the collectors know that their treasured 'arrowheads' antedate the bow."[31] In the upland areas of Louisiana—whose highest point is the 535-foot Mount Driskill—Indigenous projectile points of Clovis, Scottsbluff, and San Patrice are scattered across the hills. "Relics such as these are clear signs of prehistoric hunting Indians, marking their arrival in the area at about 10,000 or 12,000 years before the Birth of Christ."[32]

Artifacts excavated at Poverty Point, located on Bayou Macon seventeen miles west of the current path of the Mississippi near Tunica, Mississippi, reveal a center of vast trade and exchange. Discoveries include hematite, magnetite, and banded flit from Tennessee; white and gray flint from Illinois; Schistose stones and steatite from Georgia; and copper from the Great Lakes Region. There have been "thousands upon thousands" of microlithic tools found, as well as rare ceramic pottery and millions of fired-clay objects used in earth-oven cookery.[33] The people of Poverty Point spent five hundred years constructing a concentric circle of six mounds surrounding a 37-acre plaza. The largest mound is longer than 1,300 yards. Canonical mounds were erected to cover scattered ashes of cremations. A town was laid out on the concentric circles to conform with the relationships between sun and earth. "A giant effigy mound, representing a bird with outstretched wings and tail, dominated the entire community. Towering 75 feet above the flat alluvial plain, the huge structure is probably the second largest mound ever built in the eastern woodlands of America."[34] This may have been one of the earliest chiefdoms in the country.

De Soto's expedition recorded thirty sizable towns along the west side between the mouths of the Red and Arkansas Rivers. They were heavily populated and fortified. Flat-topped mounds were used as foundations for buildings. Major settlements usually contained a central nucleus: a plaza ceremonial ground bookended with a temple containing a perpetual fire at one end and the home of a chief at the other.[35] Clusters of dwellings and individual houses would be dispersed widely around the nucleus and linked by irregular footpaths. Most dwellings had a small kitchen garden. Family-controlled, cultivated fields extended a mile or more from the center of the settlement. Some towns were neighborhoods scattered through the forests and interspersed over many square miles. "The inhabitants of these towns may have been Tunica, Koroa, or ancestral Taensa, the last Louisiana tribal group closely related to the Natchez."[36]

Early Spanish maps of the Red River indicate a mound center near present-day Texarkana. There are also two well-known mounds on LSU's campus in Baton Rouge. They have been rolled on by children, parked against by football fans, and generally treated cavalierly until recently. The mounds may be the oldest known man-made structures in the America. They predate the Egyptian pyramids. Radiocarbon dating by a research team in 2022 detected ash lenses with structural components of plants and bits of burned mammal bone to date construction around eleven thousand years ago.[37] "Building of the LSU Campus Mounds shows a hiatus when climate deteriorated during the 8200 Climate Event, which defined the end of the Holocene Greenlandian Stage and the beginning of the Northgrippian Stage." Construction resumed 7,500 years ago, and construction continued on both mounds until about 6,000 years ago. LSU archaeologists speculate that scattered bands of tribes gathered there to exchange information, socialize, and perform rituals necessary for social maintenance. The oldest, Mound B, mainly consists of loess, a fine dust from glaciers in the north-central United States that can also be found along the river's east bank bluffs north of Baton Rouge up to Vicksburg, Mississippi, where the Tunica lived. It may have been used for ceremonial or cremation purposes. The younger mound, Mound A, was built from mud deposited by the Mississippi River. They stand where the Mississippi River once made a turn to the south. The Mississippi was always on the move, and mounds may have provided visible navigational points when the river was in high seasonal flood stages.

Prehistoric tribes also moved freely. Some settlements were occupied early and held for long periods; others were taken up by successive tribes. "Such places offered natural advantages like freedom from flooding, easily worked good soils, and abundant fresh water, game, fish, or salt."[38] Some lands were reoccupied after being already cleared for agriculture. Others were suitable for defense or trade. Some were associated with spiritual qualities.

There are several perennial sites along the Mississippi River, including the Houma site at present-day Angola; the settlement at Jonesville at the confluence of the Ouachita, Black, and Little Rivers; and the Marksville site, among others. Making news recently was the revelation through radiocarbon dating that a skull fragment found in 1985 at Lake Pontchartrain belonged to a woman who lived thirty-five hundred years ago.[39]

As late as 1700, Louisiana was home to several distinct native tribes. Early European observers between 1690 and 1794 reported the presence of more than thirty-two independent languages in Louisiana alone, as well as eleven languages brought by immigrant tribes being pushed out from the east. Dialects were from six major trees: Caddoan in northwestern Louisiana, Atakapa in southwestern Louisiana, Chitimacha in central South Louisiana, Tunica and Koroa in the northeast, Natchez along the Mississippi and lower Red Rivers, and Choctaw-related or Muskogean languages in southeastern Louisiana. Atakapa, Chitimacha, and Tunica were grouped as members of the Gulf stock of languages, yet to their speakers, other languages in the region would likely have sounded foreign. The Siouans, Biloxi, and Ofo moved to Louisiana after 1764 due to settlement pressures, raising the number of language families to seven.[40]

In 1682, the Quinapisa occupied a village on the right bank of the Mississippi near Hahnville and lived in several villages down toward the mouth of the Mississippi. They may have been the group that attacked La Salle's party and sent messengers to their allies, the Natchez and Koroa, to do the same. By 1700, they had joined with the Mugulasha and were allies of the Houma and Acolapissa. Shortly after 1700, the Acolapissa moved to Bayou Castine near Lake Pontchartrain to escape English and Chickasaw slave hunters. Fleeing a deadly epidemic in 1718, the

Acolapissa established a village on the Mississippi River above the new settlement of New Orleans. They eventually joined a mixed tribal group that came to be identified as Houma, among the "Petites Nations" of southern Louisiana.[41]

The Washa are believed to have originally lived in the Barataria area among many sites, including the Cote des Allemands (German Coast) post in 1739, and may have been the bison hunters reported by La Salle in a meadow below New Orleans. "The Koasati (also known as Coushatta) were driven from one site by ants and from another by rattlesnakes."[42] An eighteenth-century Bayougoula town was built on the level floodplain near the Mississippi River. "Energized" by LaSalle's journey, a host of missionary priests came to Louisiana, journaling observations while attempting to minister to Indigenous peoples.[43] Many of the French Jesuit priests lived in Indigenous settlements, detailing primary accounts that include firsthand knowledge of Taensas, Caddo, Natchez, Avoyal, and Bayougoula. Henri de Tonti, one of La Salle's lieutenants, visited the Taensa villages on Lake St. Joseph in 1690 and pushed overland to Caddoan-speaking Natchitoches tribes on the Red River. "Near the end of the 18th century, Martin Duralde, who was commandant of the Atakapa and Opelousa posts, composed a manuscript on the Chitimacha and Opelousa that is now invaluable as virtually the only source on the latter."[44] There are obvious limitations of such one-sided accounts. Some Indigenous people responded to the influx of whites by withdrawing from principal trade routes and white settlements. Others were misidentified by observers and may have mistaken for poor westerners.[45]

Excavations after Hurricane Katrina by the Federal Emergency Management Administration (FEMA) discovered pottery pieces, bones, and clay pieces of pipe dating to the late Marksville Period, 300–400 CE, at what is believed to be a midden mound at the mouth of Bayou St. John in New Orleans. The mound was subsequently used as a foundation for the French Fort St. Jean and, in 1823, as a hotel and amusement park.[46] "Though largely ornamental today, Bayou St. John was once the city's primary natural drainage outlet. It served as a major nautical access point from Lake Pontchartrain to the Mississippi River."[47] Other Indigenous settlements just prior to French colonization included the present-day French Quarter near Conti and Chartres Streets and upriver at what is now the Lower Garden District near Orange and Constance Streets. "Tchoupitoulas Street shares an origin with the 'village of the Chapitoulas' or 'river people' in Choctaw, as recorded in 1718. The Chapitoulas were one of the small groups that moved up and down the river according to trade routes and seasonal hunting in the 1600's and early 1700's."[48]

In the garden behind St. Louis Cathedral at Jackson Square, archaeologists discovered the oldest known building from the colonial period: a palmetto thatch hut built in the Indigenous style, reinforcing the theory that Native American expertise helped build the city. "Hand-built pottery, smoking pipes, trade beads, and stone hide scrapers are found in underground layers scattered throughout modern-day New Orleans."[49] Well after colonial settlement, the site behind

St. Louis Cathedral was used by Native American traders to exchange goods on market days. Native American hunters, fisherman, and herbalists supplied the French Market there well into the 1800s. "From New Orleans, natives were a ready source of trade. Furs, basketry, wild honey beeswax, and herbs were negotiable items until the 1930s."[50]

Bayou tribes modeled to the earliest white settlers how to cultivate corn, squash, potatoes, tobacco, and other Indigenous crops. The tribes, along with enslaved Africans, who were familiar with the cultivation of rice, literally saved the French.[51] "Many of the early French settlements of lower Louisiana began in Indian villages. During the frequent famines that gripped the colony, French soldiers were sent to live with Indian tribes so they would not starve to death."[52] New Orleans and Louisiana failed to prosper under French occupation. The population shrank dramatically between the original census of 1718 and the census of 1726. The 5,420 whites in the territory dwindled to only 1,952 French and German citizens (plus 276 people listed as indentured servants). The population loss was attributed to mortality rates. Then, in 1729, the Natchez and their allies revolted against the settlers, killing 10 percent of the white population.

While most of the Indigenous tribes lived along the natural levees of rivers and bayous, some like the Atakapa were coastal dwellers. They focused less on agriculture and instead gathered shellfish, fished, and hunted. "In some respects, the coastal marshes are hostile to man. The lowland is wet much of the time, the tall grasses are coarse and sharp, the swarms of deer flies and mosquitoes are nearly unbearable, and from time to time, the area is beaten savagely by the great tropical storms."[53] But it made up for these hostilities with food: mussels, clams, and oysters; and crabs, crawfish, and other crustaceans. Other foods included migratory ducks, geese, and herons, as well as mink, otter, alligator, and deer. In such a stoneless alluvial environment, shellfish also provided raw materials for containers and implements. Pottery and trade artifacts of copper, galena, and stone have also been found on the western coastal cheniers, fossil beaches named by the French that appear as low sand ridges. In Vermilion Parish near Lafayette, the Atakapa are linked to the construction of a great shell mound in the shape of an alligator 600 feet long that was visible at Grand Lake until the inundation of coastal waters from oil and gas activities.[54]

John Law's Company of the Indies, which had been granted ownership of the colony in 1717, returned Louisiana to the French crown in 1731 after fourteen years of failure. Writing from Mobile, Diron d'Artaguette described dire conditions in 1733: "Our planters and merchants here are dying of hunger, and those at New Orleans are in no better situation. Some are clamoring to return to France; others secretly run away to the Spaniards at Pensacola."[55] To the early colonialists, deltaic mud was a nuisance for the type of European knowledge they brought with them.

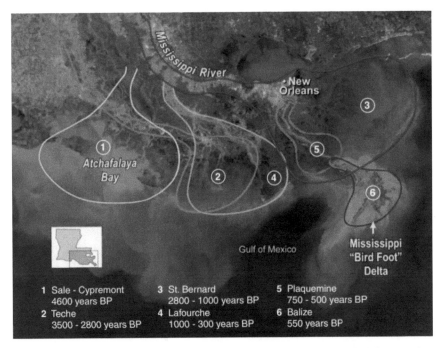

FIGURE 2. Path Lobes. Every 1,000 to 2,000 years the river made a major course change and extended new path lobes to the Gulf of Mexico, forming a new delta complex. The current Plaquemines-Modern path lobe complex, which reaches down to the Mississippi River's mouth, extends in three narrow channels that conjure the image of a claw that inspires the name, Bird's Foot Delta. Image available through Creative Commons by Angelina Freeman et al., *Water* 13, no. 11 (2021), https://doi.org/10.3390/w13111528.

GREAT SEWER

The land itself was quite unusual. As the Mississippi's mudflow reached south of Baton Rouge, it built path lobes that extended the riverine delta into the sea. The longer the ridge held the channel, the farther out the path lobe extended, until the river jumped, and its avulsion began forming a new path lobe, effectively abandoning its former channel. Current barrier islands that protect contemporary coastal communities from storm surge are remnants of former path lobes.[56] Geologists identify six major episodes of natural land construction, resulting in six distinct delta complexes that comprise sixteen separate delta lobes. "Before recent artificial levee construction, channel avulsions created a new course for the Mississippi River every 1,000 to 2,000 years." Each time this happened, a new delta complex began and the former delta slowly deteriorated. Old path lobes often become distributary bayous, which, without the Mississippi's freshwater flow, ultimately recede and subside under the counterforce of tidal surges and saltwater inundation.

The current Plaquemines-Modern path lobe complex that reaches down to the Mississippi River's mouth extends in three narrow channels that conjure the image of a claw, which inspires the name, Bird's Foot Delta. There is some disagreement among geologists as to whether this portion of the delta was already in retreat by the time of European arrival.[57] Later immigrants formed Bayou communities along former path lobes in the eighteenth, nineteenth, and twentieth centuries, including migrant Indigenous tribes pressured from the East. "Harassed tribes such as the Houma found refuge literally at Land's End, occupying the attenuated natural levees that extended toward the Gulf of Mexico."[58] Their main thoroughfares were the waterways themselves, which lie nearly even with the height of adjacent banks. Today historic civic buildings and houses face each other from opposite banks across the water on Bayou Lafourche, Bayou Terrebonne, and Grand Caillou. They are encircled by levees that lie a few leagues back on either side. These levees have become essential ramparts against encroaching water from subsiding delta marshes.

Where once the muddy river poured its sediment into its adjacent marshlands until it found new pathways to generate, western settlers demanded more control and set about building levees to protect against floods and allow for farming. With every flood that overflowed into the governed landscape, residents petitioned their local, state, and national governments for higher and stronger levees. The efforts led to a national project to control all of the nation's waterways, starting in the early decades of the nineteenth century, which pushed levee coverage up and down the nation's rivers.[59] After the great floods of 1927, the federal government enacted a massive levee program that permanently sealed off the Mississippi's sediment flow from its adjacent marshes. What had been one of the most productive wetlands in the world was effectively drained.[60]

Today, the river's sediment and water are trapped within a single channel, except for prescribed spillways maintained by the Army Corps of Engineers to regulate water flow during floods. As the largest river basin in North America, the Mississippi watershed is perhaps the most studied and controlled in the world, draining 41 percent of the continental United States: "It is not a matter of a few dams here and there. There are 40 dams on the upper Mississippi itself above St Louis. There 15 dams on the Mississippi above Yankton, South Dakota, and 21 dams on the Ohio. It's when you start to look at the tributaries to the tributaries, however, that the real picture begins to come into focus."[61]

There are more than 500 dams on the various forks of the Platte River. In the state of Kansas, there are 6,087 dams on tributaries to the Mississippi River, while the Missouri has another 5,099 and Oklahoma has 4,758. In Iowa, there are 3,340. Montana has 2,917. In the eleven states that lie entirely within the Mississippi River watershed, there are more than 30,000 dams. The total number of dams that alter the Mississippi River watershed exceeds 50,000.[62]

Within this now-entrapped water column resides an archive of the nation's detritus. Along with mountainous minerals and continental dust that erode and find their way south, we find remnants of industrial manufacturing, agriculture, and municipal effluent, along with nitrogen and synthetic chemicals from fertilizer, herbicides, and pesticides. Microbial matter, plastic litter, coal ash, concrete morsels and clay, nails, Styrofoam, industrial lubricants, and runoff from paved midwestern surfaces carry all sorts of debris from various tributaries to concoct a dystopian stew.[63] Our mud lump has grown more toxic as it moves through the contemporary industrial landscape. Appalachia footprints mingle with Illinois corn husks and swiftly bounce against man-made jetties and concrete riprap set down by the Army Corps of Engineers. Just north of St. Louis, the Mississippi no longer resembles the rather picturesque and pastoral setting of its upper reaches. As the river passes the Ohio River tributary, its color turns dark brown. An image of this confluence shows two distinct waters hesitant to commingle: diachronic strips that uneasily commune into a brown hue south of Cairo. It is said that the Missouri gives the Mississippi its volume and the Ohio its color.

Some of our congregant sediment is forced to the bottom of the riverbed where it scours against other materials in the swift rush or joins with other fragments and castoffs. Some lumpy ooze will veer off just north of Baton Rouge through the Old River Control complex gates, which are situated in an old oxbow path of the Mississippi that nearly touches the Red River. Prior to the fifteenth century, the Red River and the Mississippi River flowed parallel to each other to the Gulf of Mexico. Just as Europeans were arriving in Louisiana, the Mississippi created a natural oxbow that briefly intersected into the Red River. This created a confluence with the Red River that drew the Upper Red into a tributary to the Mississippi and the lower portion of the Red into the Atchafalaya distributary. The Atchafalaya/Red complex was also congested by a massive raft of logs referred to as the Great Red River Raft. As McPhee describes it, "The raft was so compact that El Camino Real, the Spanish trail coming in from Texas, crossed the Atchafalaya near its head, and cattle being driven toward the Mississippi walked across the logs."[64]

During the riverboat period of the early nineteenth century, vessels took as long as twenty hours to travel the oxbow, called Turnbull's Bend, to advance one mile as the crow flies. In 1831, an enterprising riverboat captain named Henry Shreve proposed to quicken the journey by cutting a mile-long canal from the Mississippi to the Upper Red River complex across a narrow tuft that shaved off nineteen miles from the oxbow route. It also avoided the massive logjam in the Lower Red that blocked passage. Shreve's cut would later be reframed as an act of immense hubris that contributed to a permanent "crevasse" of the Mississippi River into the Atchafalaya distributary.

Since the Atchafalaya system was deeper (and closer to the sea), the Mississippi River would periodically flood during high water through Shreve's cut into the Atchafalaya distributary. "Snag boats worked on it, and an attempt was made to clear it with fire. The flood of 1863 apparently broke it open, and at once the Atchafalaya began to widen and deepen, thus increasing its draw on the Mississippi."[65] Shreve's cut at Turnbull Bend had two effects: it opened the Red River Valley to steamboat commerce, and it created the unintended consequence of promoting a new path lobe by the Mississippi into the Atchafalaya. Although this is conventionally accepted, it has likewise been challenged. The river may have jumped without Shreve's nudge. "Since most of the problems at Old River arose after Captain Henry Shreve constructed his cutoff in the area in 1831, it has often been argued that Shreve was responsible for the difficulties."[66]

As early as 1829, Samuel Cummings of the *Western Pilot* commented that the river was rapidly wearing away the neck of land in the bend, which was then only 200 to 300 yards across. "Islands No. 119, No. 120, and No. 121 (called the Three Sisters) had already been completely washed away."[67] By 1900, river gauges indicated that 10 percent of the Mississippi River was flowing into the Atchafalaya. With each flood, the gap widened. By 1930, about 20 percent of the Mississippi flowed through the crevasse, and by 1950, 30 percent flowed. At the time, the Corps estimated that by 1970 the entire Mississippi would be irreversibly captured by the Atchafalaya distributary, which would turn New Orleans and Baton Rouge into bayou cities. Such a dramatic shift in course also had the potential to spoil the drinking water for the 1.5 million people that relied on the Mississippi River Delta. Meanwhile, a corridor of petrochemical plants had grown up in the second half of the twentieth century between Baton Rouge and New Orleans: "As a result of settlement patterns, this reach of the Mississippi had long been known as 'the German coast,' and now, with B. F. Goodrich, E. I. du Pont, Union Carbide, Reynolds Metals, Shell, Mobil, Texaco, Exxon, Monsanto, Uniroyal, Georgia-Pacific, Hydrocarbon Industries, Vulcan Materials, Nalco Chemical, Freeport Chemical, Dow Chemical, Allied Chemical, Stauffer Chemical, Hooker Chemicals, Rubicon Chemicals, and American Petrofina—with an infrastructural concentration equaled in few other places—it was often called 'the American Ruhr.'"[68]

Just as colloquially, this stretch is commonly known for the distinction of having the highest emission concentration of carcinogens in the United States. The industrial plant operators, who purchased large parcels from sugar plantations, had moved there to take advantage of the river's deep-draft access to fresh water—and little regulatory oversight. "They would not and could not, linger beside a tidal creek. For nature to take its course was simply unthinkable."[69] Congress authorized the Old River Control Structure, a dam that opened in 1963.

The Corps continually monitors water levels of the Mississippi and Red (and Ouachita) Rivers to apportion 30 percent of their combined volume down the Atchafalaya to prevent the Mississippi from permanently changing courses and claiming the Atchafalaya. Through more than fifty years of existence, the Old River Control Structure has nearly been lost twice and now comprises five man-made canals, including two control channel spillways, a lock from the Red River to the Mississippi, and a hydroelectric dam, in addition to the nearby Morganza Spillway.[70] The entire footprint sprawls over 250,000 acres of easements within the project area, having displaced several historic towns and farmsteads to keep the Mississippi from repeating the disaster of the great flood of 1927. Such regimes of technopolitics to control nature are often the children of earlier failures.[71]

Old River Control is part of the ongoing Mississippi River and Tributaries Project, known as Project FLOOD, that was authorized after the great flood of 1927. Project FLOOD is the largest flood control project in the world, intended to provide protection for the 36,000-square-mile lower Mississippi River Valley of 1.5 million homes, 33,000 farms, and 4 million people. Its interventions include meander cutoffs, jetties, diversion spillways, and a levee system 2,203 miles long, including 1,607 miles of levees along the Mississippi River itself, and 596 miles of levees along the south banks of the Arkansas and Red Rivers and the boundary of the Atchafalaya Basin.[72] This protection required social sacrifices in the form of expropriated farmlands and displaced minority communities to allow for large spillway easements for the controlled floods.[73]

Here we say good-bye to some of our mud as it forks to the west through the sluice gates and canal of Old River Control into the Atchafalaya River and Swamp, which is the largest wetland swamp in the United States. The rest of our mud and matter will continue down the Mississippi's main stem past Baton Rouge as it enters Cancer Alley, the 85-mile-long cluster of plants in the "River Parishes" of Ascension, Iberville, St. James, St. John the Baptist, and St. Charles. This corridor continues to open its banks to newer and more sophisticated plants that produce industrial materials, fertilizers, and lubricants impervious to biodegradation. Their production is imminently harmful to *fence line* neighbors, who are disproportionately African American and low income. Lately, the corridor has been touted by the Louisiana governor and fossil fuel advocates for its potential to host Industrial carbon capture projects. The state is actively recruiting more plants, which are supercharged by a boost in federal subsidies under the Biden administration and a ten-year state property tax exemption that shields heavy industry operators from paying hundreds of millions of dollars a year.[74]

Gaining in toxicity, our mud lump continues toward New Orleans. It nears the site of four former plantations, Delhommer, Roseland, Myrtle Land, and Hermitage, which became the townships of Sellers and Montz. Between the two former townships sits the Bonnet Carré Spillway, which is a huge drainage canal that connects the Mississippi River to Lake Pontchartrain. It was constructed in 1931 as the last line of flood defense for New Orleans. The Army Corps of Engineers opens it when the Mississippi exceeds the designated 17-foot flood stage. The spillway was built atop two African American cemeteries for people enslaved on the four plantations and their descendants. The cemeteries were still being used when the Army Corps of Engineers began excavation for the spillway. The Corps was supposed to relocate the cemeteries, but there is no record of it happening.[75]

The Bonnet Carré Spillway consists of 350 bays of giant cypress planks along a mile-long trellis. They must be opened individually by hard-hat crews using cranes to remove the planks one by one. The open bays allow the river to flood into the marshes and brackish tidal waters of Lake Pontchartrain. In ninety years, the spillway has been opened thirteen times. But the frequency has dramatically increased in recent years. In 2019, the river broke the record set in 1927 for the most continuous days at flood stage. For seven months, the Mississippi towered over the rooftops of New Orleans.

Our lump passes out of Cancer Alley into the western edge of the New Orleans metropolitan area, which is demarcated by the $14.5 billion federal levee system rebuilt by the Army Corps of Engineers after its levees failed in two dozen spots during Hurricane Katrina. Much of the land within the levee system sits below sea level, which means that rainwater must be pumped up and out by giant pumping stations. Suburban areas of metro New Orleans were built on drained marshland in the twentieth century. However, the oldest historic settlements follow the snaking embankment of the river, which offered the highest alluvial ridge, at around sea level.

As our toxic lump enters the New Orleans city limits, it will whip around and flow due north at the French Quarter—site of the original colonial settlement in 1718. La Nouvelle-Orléans was chosen by the French Canadian naval officer, Jean-Baptiste Le Moyne, Sieur de Bienville, and his brother, Pierre Le Moyne d'Iberville, who believed that the settlement would provide the French crown with access to the Gulf of Mexico in order to stake a claim deep into the interior of a contested continent. The site was accessible not only by the great river but also by a short portage to Lake Pontchartrain used as a fishing, hunting, and gathering point for regional Indigenous tribes. Before the establishment of New Orleans, there were corn feast celebrations at the bayou's terminus behind today's French Quarter known as Congo Square.[76]

Just past New Orleans, the lump passes the Chalmette Battlefield where British forces attempted to lay siege to the city in 1812. The curated battlefield, which is a popular tourist attraction, was actually "reclaimed" by the National Park Service in 1968. The service evicted people living there in an African American "freetown" called Fazendeville that consisted of thirty families in 1964. After the Civil War, the plantation of Ignacio de Lino de Chalmet was shuttered and eventually purchased in 1870 by a free man of color, Jean-Pierre Fazende, who divided the land into thirty-three lots and sold it off. The community of Fazendeville was referred to as "the Village" by its residents. It had grocery stores, a baseball diamond, two churches, a one-room school, and a few dozen homes. Today the community is entirely erased. It is memorialized only by a slight depression in the landscape and a line of Louisiana irises that some believe were planted by the residents who were forcibly relocated to make way for the national park to commemorate the 150th anniversary of the Battle of New Orleans.[77]

As our mud lump flows by the city and makes its final twist at the Ninth Ward, it will veer dramatically southeast toward the Head of Passes just above the Bird's Foot Delta 90 nautical miles from New Orleans.[78] Along the final stretch of river, some of our material will collect in sandbars, where it is dredged by the Army Corps of Engineers to maintain the channel, reflecting a recent political push to deepen the river from 45 feet to 50 feet for the larger, Panamax supertankers that can now traverse the Panama Canal. In 2016, Panamanian officials opened a third lock that increased the 39.5-foot draft for container ships to a 50-foot draft. The new lock set off a proverbial arms race among US port cities to capture deeper cargo ships that had previously sailed around the tip of South America and avoided the Gulf of Mexico and the Caribbean.[79] The Louisiana Port Association hired an LSU economist to write a report about the importance of the shipping industry to Louisiana's economy by providing "one in five" Louisiana jobs. This report has become a standard talking point and rationale for any port expansion.[80]

A sharp southwestern turn in the river—known as English Turn for the famous bluff by Bienville—creates a protective elbow peninsula that shields New Orleans from a relatively direct route to the Gulf of Mexico 75 river miles away.[81] Today the peninsula houses a few large estate homes and an upscale neighborhood, as well as the Audubon Institute Species Survival Center, which is leased from the US Coast Guard to care for sick animals. Along the river, anchored bulk cargo tankers await an open berth at the Port of New Orleans. A small clearing of trees provides a nice casting spot for small groups of fishermen hoping for catfish. Blue herons, white egrets, and turkey vultures perch along the batture forest. They are more visible these days in a canopy that has been severely thinned in successive storms: Hurricanes Harvey in 2017, Barry in 2019, Zeta in 2020, and Ida in 2021. Across the wide breadth of the river, the thirteen-deck Carnival Cruise liner *Valor* speeds downriver at a fifteen-knot clip, towering above a fully loaded tanker crawling its way upriver.[82] In the grass that edges up to the levee's concrete armor, remnants of plastic cups, buckets, and Styrofoam containers have washed ashore. Seagulls and insects chirp through the industrial soundscape. The fishermen tend their rods after casting, place them against stands, and smoke a cigarette or peer down at their phones. Just upstream, excavators and dump trucks work a sand quarry, Parish Sand LLC, which uses a river water intake method from pipes, which when dried leaves behind sand and sediment.

From the water, some of our dredged sand and slurry will be sucked through large pipes and directed into scattered clumps of vegetation to try to reclaim Louisiana's disappearing barrier islands and marshes at a cost of about $3.7 million per acre.[83] Natural crevasses are also opening and widening above the Bird's Foot Delta, which is setting off tension between the Corps of Engineers, which wants to seal them back in order to maintain a scouring effect for shipping, and state coastal interests, which would like the openings to demonstrate if sediment diversions can rebuild adjacent marsh there.[84] Redirecting sediment upstream also spells trouble for the Bird's Foot Delta, which is sinking despite constant dredging by the Corps and using "beneficial material" to reinforce its edges and interior at the forest.

Most of the lump will plume out in brown wispy tendrils into the green and cobalt waters of the Gulf of Mexico. From high in the sky, its coherence appears sanguine, but on the surface, we discover an aquatic dead zone for hundreds of square miles. This procession of nutrient-rich material is digested by blooms of algae that quickly die and decompose in a bacterial metabolism that consumes the water's surface oxygen. A resulting hypoxia suffocates fish and shrimp and kills off stationary species like oysters. It worsens in the summer months as the warmer river water glides over colder, dense salt water below. Less oxygenated water is trapped at lower depths. The zone grows as human-activated climate change amplifies the natural fluxes of this area. According to scientific measurements, this hypoxic zone equals the size of Vermont.[85] As we try to frame what this entire journey and changing landscape implies for the land and the inhabitants, let us return to the modern imagination that took hold of this landscape.

CONCLUSION: BUILDING A MODERN SETTLEMENT

The project to build a modern settlement in a delta required erasing the ubiquitous mud that created it. It required a constant redoubling of interventions that paradoxically increased its vulnerability. For all of its existence, the communities of the Mississippi River Delta—including New Orleans—have lived in a state of hypervigilance of their surroundings: the Mississippi River on its front side and the swamps to its rear. The river was historically framed as the region's greatest asset if not also its threat. Urban footprints relied on intensive drainage, levees and labor, pumps, and money to erase mud altogether. This battle of erasure historically sustained the geographic, administrative, social, and economic structures of the region.

As interests compete over a shrinking footprint, mud is pulled from the floor of the river mechanically by dredgers to allow shipping interests to prosper. It is pumped into the marsh. It is used to bolster barrier islands. It is trapped in spillway gates. It even lives in the form of wetland mitigation credits that can be purchased from private mitigation banks that own land being reforested as "offsets" to allow industrial construction in vulnerable marsh. Mud builds estuaries by holding vegetation together. It is used in oil and gas drilling to cool the drill bits and stabilize bore holes. It produces value for contractors that are studying it, capturing it, and moving it. It is rendered and represented in scientific reports, software modeling, and maps. In some cases, it is produced through sewerage and municipal effluence.

It has been the mud that sustained the people of New Orleans and Mississippi River Delta, despite their historic aversion to it. "Mud is, or was, the essential building block of the Mississippi Valley. Mud was the substance that held together all the pieces and components—animal, vegetable, human of the floodplains' wetland ecology."[86] Mud is today desperately needed. But there is simply not enough to replenish the subsiding boundary of Louisiana as hurricanes roll up marshland too quickly for the ecology to recover and subdivisions clear-cut the remaining bottomland forests to accommodate inland coastal migration. As the marsh transforms into open water, inland the root systems of hardwood are being stressed by growing salinity and saturated bottomlands.

By examining how delta mud was written about, defined, discussed, and otherwise represented, we can track attitudes about mud through colonial and racializing discourses that persist today. This may allow us to disentangle some of the associations at work that conceal troubling values tied to coastal interventions. By reconsidering mud as a primary building block, rather than an afterthought, we come to realize the story of mud is also the story of the people of this delta.

Interlude

On the Mermantau

Early morning in the Atchafalaya Swamp, fragmented colors ricochet off contours of knobby-kneed cypress roots. A great heron perches on a branch. The swamp's mud reeks of sulfuric eggs. Its dank residue easily stains your pant legs. Fishing camps loiter on graveyards of pine needles. Wood shakes and twigs surround fire pits. There is no hill or vantage on which to stand over the lush, buzzing scene. Just a deep expanse and a skyline of trees.

Growing up, my grandparents' fishing camp even then felt dated. A mounted bass, perched over the kitchen entrance, pointed to another time. A finger cut on a fishing hook must be cleaned with soap. The Mermentau River south of Abbeville is technically a bayou. It has little flow. It's wide enough to swim, fish, and even water ski. The water is espresso colored. You avoid touching the bottom, afraid of snake holes and alligator nests and soft mud that oozes through your toes. The water's surface plays tricks on your young imagination. Little heads surface around you. Bubbles perhaps. Turtles? Snakes? You splash at them to calm yourself until the ski boat returns. You hear a splash in the distance from the water's edge. "Water moccasins," says Dad, "falling from the trees." Every splash? You hear splashes constantly.

Torso-sized spider webs stretch across tree limbs. Suddenly, you realize, you are surrounded by colonies of life, and potential threats. Your older sister gets to drive the boat. You want to drive the boat. Your uncle that night feeds you stories about alligators outrunning children. "Faster than dogs," he says. You don't believe him but still wonder. Maybe you'll get to see an alligator.

2

——

Muddy Foil

Let us chart the unbroken connection between Extractive Thinking and the historically racialized project of eradicating swamps, wetlands, and mud. While the Mississippi River enjoys a rich lexicon of cultural signification, its mud indexes a very different legacy. Newspaper articles, technical reports, and other accounts document attitudes of fear about and aversion to the undeveloped swampland that surrounded New Orleans. Dark, muddy forests and wetlands were considered sites of miasmic disease, lawlessness, and a hindrance to moral conduct.

These accounts eerily paralleled attitudes toward the nonwhite human individuals who filled such spaces. Mud was disqualified. It was used in metaphors for racial miscegenation. Mud was a protean material in the act of becoming something else: not quite water, not quite land; anathema to the modern project of categorization and enclosure. In his exploration of power, Michel Foucault challenges us to analyze bourgeois power by the way in which it was applied through tools and tactics of domination. What if we applied those concerns to the way in which the landscape itself was reshaped to reproduce racial formations?[1] For example, the forces of Extractive Thinking embedded in the built environment of dams, levees, and spillways that sealed the river from its adjacent marshes are bound up in an older, more troubling legacy. Plantation capitalism drove demands for enslaved labor to build the very levees that drained the swamps for the fields themselves. The plantation economy generated further demand for enslaved bodies, who built levees up and down the Mississippi River.

Waterways were also dredged with enslaved labor, which allowed for the transportation of cotton and sugar produced on former wetlands to markets in New Orleans. The river created the delta. But it was white supremacy that brought enslaved Africans and Americans to the delta to "clear it and tame it and transform it into an empire."[2] When levees broke during the eighteenth and nineteenth centuries, their repairs were often made by the enslaved and, in the early twentieth century, by conscripted Black laborers, who at times were forced to use their own bodies to plug crevices. During the flood of 1912, the *New York Times* reported from Greenville, Mississippi, at Miller's Bend that authorities had exhausted all their sandbags and instead used bodies: "A young engineer in charge had a brilliant idea and proceeded to put it in execution. Calling to several hundred negroes, who were standing idle, he ordered them to lie down on top of the levee and as close together as possible. The young men obeyed, and although spray frequently dashed over them, they prevented the overflow that might have developed into an ugly crevasse. For an hour and a half this lasted, the negroes uncomplainingly sticking to their posts until the additional sandbags arrived."[3]

Slavery and racial power are not only embedded in the levees, plantation fields, and layout of the Mississippi River but also in the urban centers that it financed, like New Orleans. "The echo of enslavement is everywhere," writes Clint Smith. "It is in the detailed architecture of some of the city's oldest buildings, sculpted by enslaved hands. It is in the levees, originally built by enslaved labor."[4]

ÎLE D'ORLÉANS

New Orleans has little reason to exist but for the river and the political economy it supported. Its mushy topsoil and stagnant rain pools helped characterize the city as floating land: shaking prairie, or "la prairie tremblant," by the French.[5] Its paludal, or marshy, environment is "half-land and half water" composed of organic to highly organic sediments deposited there.[6]

Early settlement patterns snaked along the alluvial ridges created by the river, while areas in the soggy "back of town" were sparsely populated.[7] From its very beginnings, the city's mud shocked arriving travelers, who were described as unanimous in their condemnation of the unpaved streets, which, though well laid out, were little more than muddy canals.[8] The streets were 37 feet wide and lined with ditches to carry off seepage from the river levee. Open, crisscrossed ditches, when flooded, functioned by the "curious" phenomenon of draining water and refuse of the city away from the river toward the lower-lying "back-a-town" cypress swamps. These back swamps were described as a muddy "gruel" of water and organic matter. "Slop and garbage thrown in the gutters" created a stench that could only be expelled by flushing rains. "The blocks after a hard rain were completely surrounded by water, and as a consequence, came to be called islets."[9] Drainage alleviated flooding in the highest areas, which were along the river and canals. A visiting Captain Hamilton wrote in 1833 that after a rain, the center of the street was at least a foot thick in mud. "The only sewers," he reported, "were open drains clogged with garbage, refuse and human waste, euphemistically termed night soils."[10]

In correspondence during the early settlement period, the local Bishop de Luxembourg's requests for supplies to the New Orleans Mission illuminated the muddy reality of making a go in the delta. He describes the inadequate supplies to the friars and "a diet of a little boar, a half-pound of bread and a quarter liter of wine" after supplying for mass. "The fatigue we endure running night and day to visit the sick and carry the sacraments to them, generally in mud knee deep, does not accord with such scanty nourishment," the bishop wrote.[11]

In what contemporary geographers have framed as either New Orleans's "inevitable"[12] founding or "Bienville's dilemma,"[13] the colonists were plagued by the very mud and water on which the city's strategic value depended. The first levee around New Orleans was ordered by Governor Bienville in 1719, the year after the settlement was founded. Well before Louisiana and New Orleans's purchase by the United States, the French engineer Vitrac de La Tour had understood that the new settlement was prone to periodic flooding, and he opposed the chosen location. Bienville, however, overruled La Tour's objection and had the engineer design a 5,400-foot-long and 18-foot-wide earthen embankment along the Mississippi, completed in 1727, to protect the city from seasonal floods. The levee stood 3 feet high and doubled as a roadway. Three years later, New Orleans was flattened by a hurricane.[14]

While historical accounts overtly document the story in terms of a struggle against water, the story of mud haunts them. The river seemed to beckon settlement as the mud foiled it. Mud stymied efforts to govern a rational landscape and harvest the bounty that the river promised. Muddy streets. Muddy clothes. Surrounding swamps blamed for diseases. New Orleans's infamous mud motivated massive drainage programs to develop land more suitable for cypress tupelo than concert halls. Persistently muddy roads and dank puddles exasperated ordinary folks and building experts alike, among them Benjamin Latrobe, the first formally trained American architect and designer of the US Capitol. "Mud, mud, mud," Latrobe sighed in 1819. "This is a floating city, floating below the surface of the water on a bed of mud."[15] Western anthropogenic practices treated mud as a nuisance to be removed from the river channel, stacked along the banks in levees, and drained from behind the levees for plantation and urban development. Mud was the unwelcome interloper in the modern imaginary to stabilize the land from the river.

While the river made the city famous, the mud gave New Orleans its sense of fecundity. Backwater swamps behind the French Quarter attracted gatherings of enslaved Africans, African Americans, enslaved people from the Caribbean, free people of color, and even Native Americans. The sensual entanglement of New World colonization at sites like Bayou St. John and Congo Square through rhythmic calls, songs, dances, and orchestration helped create the art forms jazz and blues. "African-derived habanera rhythm and its derivatives, found in the most popular Creole slave songs and the correlating dances of Congo Square, are also found at the core of early New Orleans jazz compositions, second line or parade beat, jazz funeral music, and Mardi Gras Indians chants and rhythms," writes Freddi Williams Evans.[16] It is also where the distasteful caricatures of the blackface minstrel were popularized.[17]

Some archaeologists theorize that vessel fragments unearthed at St. Anthony's Garden behind the St. Louis Cathedral in the French Quarter were used to add traction to the muddy streets, since there was no natural gravel source. Early nineteenth-century New Orleans was a place of intersecting bayous and rivulets that flowed to and from the river and adjacent lakes, depending on seasonal levels. Today most are filled. Richard Campanella, a New Orleans geographer, writes that many of the so-called lost bayous of New Orleans provided natural ridges that were used as thoroughfares. There was Petit Bayou, renamed Pequeño Bayou de la Cruz by the Spanish. There was Bayou Gueno and Bayou Au Lavoir, which was used for washing clothes. There were main throughways known by natives as Bayou Coupicatcha or by French settlers as Bayou Métairie for its small tenant farms. Nearby Bayou Chantilly (later Gentilly) was named for an estate outside of Paris. It discharged into the wild marshes of Bayou Sauvage, which today is part of a drainage system on the city's eastern flank. These waterways and their haunted connections to the Mississippi were responsible for the stitched alluvial ridges and sinks of New Orleans. "Because distributaries deposited river-borne sediment, they built up ridges, or natural levees, along their banks, which were used as roads by early inhabitants."[18]

The Bayou Metairie/Gentilly ridge often impounded watery tributaries behind them, which drained into the midcity lowlands between the bayous and the river's uptown alluvial ridge. Various other small rivulets were interlaced throughout New Orleans. As the city developed, early French settlers regarded the wetlands of willow trees and cypress swamps as malaria-prone thickets to be transformed for settlement and economic viability.[19] Early houses were set on pillars with ground-floor cellars. Sidewalks were elevated and wooden. Called banquettes, they were often uneven and beset by detours around standing water. "Walking was an adventure. On more than one occasion high-born ladies went to balls with their skirts lifted high and their party shoes and stockings in their hands."[20] Alexis de Tocqueville noted the ubiquity of mud: "Fine houses, huts; streets muddy and unpaved."[21] The *New Orleans City Guide* produced by the WPA in the 1930s stated that it was a wonder New Orleans existed at all, with the "soggy nature of the subsoil, the low elevation of the city, climatic conditions favorable to malignant diseases, and danger of Mississippi River flood waters."[22] The city's unkempt conditions were attributed to everything from open sewers to the indolence of European creoles. The word *creole* itself is a slippery registry of in-betweenness. The *City Guide*'s opening pages cite a cautionary nineteenth-century minstrel.

HAVE you ever been in New Orleans? If not, you'd better go. It's a nation of a queer place; day and night a show! Frenchmen, Spaniards, West Indians, Creoles, Mustees,

Yankees, Kentuckians, Tennesseans, lawyers and trustees, Negroes in purple and fine linen, and slaves in rags and chains. Ships, arks, steamboats, robbers, pirates, alligators, Assassins, gamblers, drunkards, and cotton speculators; Sailors, soldiers, pretty girls, and ugly fortune-tellers; Pimps, imps, shrimps, and all sorts of dirty fellows; A progeny of all colors, an infernal motley crew; Yellow fever in February, muddy streets all the year; Many things to hope for, and a devilish sight to fear![23]

These words, attributed to a Colonel Creecy in the 1830s, reflected a prevailing trope of the city as not only a place of filth, but also disrepute. New Orleans was considered a risky locale of unsavory characters and pestilence—a reputation that preoccupied authorities worried about investment and commerce. Cleaning up the sources of disrepute and disease became a perennial vocation. To authorities, the problem originated from the land itself. Urban improvement focused on conquering the swamps through drainage, circulation, and enclosure. Calls for drainage were laced with public health imperatives, particularly concerns with regular summer bouts of yellow fever, among other scourges. Perennial outbreaks of disease in the eighteenth and nineteenth centuries plagued New Orleans at a level that was said to be twice that of other large urban areas.[24]

THREATENING PRESENCE

Aversion to mud has a long history in Western sensibility. Muddy wetlands with their stench, heat, and lack of solidity challenged the very foundation of Western Enlightenment and earlier logics of pre-Christian Hellenic society. As early as the fifth century BCE, we find recorded abjection to mud that disparages terrains that are uncultivated, difficult to pass, full of malaria and death and a certain ambiguity. Hippocratic writings described unhealthy waters as still and bilious. "Turbid stagnant water in marshes and swamps are hot, thick and evil smelling in summer because of their stagnation and failure to flow."[25] Pliny the Elder in the first century condemned stagnant, sluggish waters that he contrasts with beneficial running water cleansed by the "agitation" of the current. "Wholesome waters should also be without taste or smell."[26] Wetlands in their oozy, liminal materiality challenged rationalized configurations of the world because they were neither land nor water. Such mud resisted categories. It was a material in process of becoming the other. This protean threat, which was thought to house sickness and the monstrous, was later mapped onto discourses that reflected the slippery abyss of the human unconscious—a trope that emerged during the 1850s.[27]

In the colonial imagination, mud was often gendered as the feminized body or racialized in the dark jungles of Africa. In E. M. Forster's *African Queen*, the swamps are described as a dreary, marshy amphibious country, "half black mud and half water," neither solid nor liquid, not light or dark. "Undoubtedly the worst feature of the swamp was the awful smell of rotting vegetation that hung about it, which was at times positively overpowering, and the malarious exhalations that accompanied it, which we were of course blighted to breath."[28] According to Rodney Giblett, the slimy composition of the swamp is what makes it an object of horror that won't "sit still as some sort of fixed and static mediator."[29] It lurks in the "murky" edges between water and land.[30] It is at this conjuncture where utopian imaginaries battled folklore using the tools of technology and science. Wetlands were home to Western literature's famous ogres. Grendel, along with his mother, lived in a perilous marsh where the mountain stream goes underneath the mists of the cliff. A wanderer of the marsh, Grendel was guardian to Moors and alien spirits. In *Paradise Lost*, Milton's Satan is a swamp serpent and marsh monster. The swamp, like Satan, trespasses on every domain. In Dante's fifth circle of hell, the 'sullen souls' are stuck in the slime.[31] It is into the primeval darkness of the swamps that Marlow must venture to rescue the dissembling Colonel Kurtz. He describes his journey inland where "the savagery, the utter savagery," had closed around him.[32] One is never alone in the swamps; the foreboding of otherness watches from the impenetrable thicket. Swamps and wetlands buzz with nonhuman life. Louisiana's Honey Island Swamp Monster as well as the Loup Garou werewolf, supposedly inherited from France, found their way into local folklore and popular music. The "rugarou," a variation of the Cajun French version, were known as skin-walking, shape-shifting, half-humans that haunted Louisiana swamps.[33]

Conquest of wetlands by drainage was consequently framed in terms of security—if not imperial conquest. Benito Mussolini, who drained the "never-ending fen" of the malarial marshes, later bragged that his two main achievements were that he made the trains run on time and drained the Pontine Marshes.[34] Drainage, likewise, rendered land profitable for monocultivation. John Locke argued that uncultivated lands should be available for seizure, which was a rationale used by American colonists to dispossess Indigenous peoples from their ancestral lands in the seventeenth through nineteenth centuries.[35] "Some have argued that the book of Genesis still persuades many, convincing Americans that God has given them domination over nature, empowering Americans to lay waste to nature to transform resources into consumer goods."[36] We see such justification in the words of then-President Donald Trump channeling the nineteenth-century discourse of "Manifest Destiny" when speaking at a 2018 Naval Academy commencement: "Our ancestors tamed a continent. We are not going to apologize for America."[37]

Historically, wetlands were also beneficial to localized resistance, which used the shadowy thicket to stage ambushes. Guerrillas during the Revolutionary War were called Swamp Foxes. The muddy swamps of the Chalmette battlefield just downriver from New Orleans aided Andrew Jackson's forces against the British landing in the 1815 Battle of New Orleans. In the Seminole Wars, the Florida swamps were described as taking the sunshine from a man's life: "Cypress knees, mangrove roots, and saw grass tortured the foot soldier. Too much water, and the lack of water, made his life a torment. There was marching in water from ankle to armpit deep, hour after hours, with no chance to dry off, not even light."[38]

THE UNFOLDING HISTORY OF MARRONAGE

William Styron's fictional confessions of Nat Turner muse about the stronghold of the swamps, which were "profusely supplied with game and fish and springs of sweet water—all in all hospitable enough a place for a group of adventurous, hardy runaways to live there indefinitely, walled up in its green luxuriant fastness beyond the pursuit of white men."[39] Harriet Beecher Stowe's runaway slave, Dred, takes refuge in the Great Dismal Swamp, which becomes a symbol for madness. She describes its "goblin growth[;] . . . all sorts of vegetable monsters stretch their weird, fantastic forms along its shadows."[40] Archaeologists have unearthed evidence that self-liberated enslaved people persevered for generations in the Great Dismal Swamp, evading capture by slavers and allying with Native Americans, themselves fleeing the colonial frontier and forced resettlement, from at least 1680 to the Civil War nearly two centuries later.[41] Similarly, the protective geography of Louisiana allowed those wishing to avoid authorities to move freely among interconnected waterways and forests. The lands behind wealthy manors and plantations transitioned into cypress swamps, known as *la cipière*, where much activity was unsupervised. "The lands on and behind the estates afforded excellent, nearby refuge to runaway slaves," writes Gwendolyn Midlo Hall. "Neither master nor overseer was eager to venture into the swamps."[42]

These marginal, untamed spaces and their shadowy inhabitants posed a direct challenge to planters and overseers, who attempted to impose a hard line of separation between their sphere of control and wilderness. Despite laws to keep enslaved people from interacting with other households, those who had escaped the plantation, known as maroons, regularly met with enslaved people from different plantations and maintained secret networks along rivers and bayous. Maroons also married people who were enslaved on plantations. They might return to the cabin of a loved one for food, putting bay leaves on their shoes or tracking through fresh manure to throw off the scent of dogs.[43]

Such liminal spaces between water and land were home to a still-unfolding history of Black residents in what might otherwise appear as unremarkable territory of mud and marsh. By coexisting with these ecologies, they were able to use the wilderness as a defense. "Each time a maroon community claimed space in a landscape under the nominal control of an early modern state, it established a geographical 'maroon landscape,'" writes the historian Marcus Nevius.[44] A maroon landscape ranged from borderlands proximate to plantation societies to remote hinterlands to which rebels escaped to fully repudiate enslavement. By the American Revolution in 1775, maroons of Louisiana occupied the areas between the mouth of the Mississippi River and New Orleans, known as the Bas du Fleuve.[45]

They subsisted in the tidal wetlands near the Gulf of Mexico, rich in fish, shellfish, and game.[46] Eventually, maroon territory stretched up and down the Mississippi River—from St. John the Baptist and St. Charles Parishes immediately upriver from New Orleans to downriver from the English Turn through Lake Borgne—where the swamps were nearly impenetrable. Until the Civil War, there were thousands of people who joined maroon encampments in the vicinity around New Orleans from as far north as Pointe Coupee Parish north of Baton Rouge.[47]

Spanish authorities of late eighteenth-century Louisiana were deeply concerned by the military strength of maroon settlements, particularly the bands of resistance in "Gaillardeland," which were uncharted swamps in present-day St. Bernard Parish downriver from New Orleans.[48] These territories were home to the so-called San Malo Maroons. Largely self-sufficient, they cut and delivered cypress logs to mill owners for cash. They fished and hunted. They grew beans, corn, and herbs that were sold in street markets in New Orleans. They were armed with muskets whose shot and powder were purchased in New Orleans. But they were also fluid communities that had to navigate deep waterways that were home to alligators, snakes, mosquitoes, and other dangerous wildlife. Challenging as they were, the wetlands along Lake Borgne provided a natural barrier of protection, away from colonial authorities and the grip of racial slavery.

The language of early modern observers, particularly in North America, overemphasized the perceived threat posed by maroons. Maroon settlements were depicted in New Orleans newspapers as sources of danger and ambush.[49] "Slave hunters and other pursuers were slain during small-scale raids; but large anti-maroon operations were noticeably one-sided when it came to the loss of human life." Despite the fear they provoked, maroons did not inflict much bodily harm on the white population. Within 250 years, probably fewer than 150 whites were slain during revolts.[50] Newspaper accounts describe efforts by authorities to suppress maroon activity well into the nineteenth century. The largest maroon communities attracted the most attention and generated the most pervasive fears among colonials. Accordingly, scholars, seeking to explain maroon community formation, regularly studied the largest maroon communities as evidence of grand marronage, or permanent removal from plantations to settlements. By contrast, short-term flight undertaken by enslaved individuals or small groups came to be known as petit marronage. Nevius writes, "As Thompson observed in his 2006 book, *Flight to Freedom*, these studies have generally turned earlier readings of slave resistance on their head to reveal that, by comparison with the outbreak of outright rebellion, arson, poisoning, and other forms of resistance, marronage was the most pervasive action that enslaved people undertook to be free."[51]

Punishment for leaving the plantation without a transit pass could be severe. An extended grand marronage was determined by duration, distance traveled, and the number of prior offenses.[52] According to the French Code Noir of 1724, which regulated interaction between whites (*blancs*) and blacks (*noirs*), sentences

FIGURE 3. Marronage. A drawing of an imagined self-liberated maroon encampment in Louisiana published in 1878 in *Harper's Weekly*. Muddy swamps and thick forests provided natural refuge to self-liberated individuals throughout the southern United States.
Image courtesy of the Historic New Orleans Collection, 1982.54.1.

for a one-month marronage included cutting ears off and brandings of fleurs-de-lis on the shoulder for a first offense, hamstringing and fleurs-de-lis brands on the other shoulder for a second offense, and death for a third offense. Penalties for free people of color who harbored runaways ranged from paying masters of the runaways "30 libbers" for each day they were gone or, if they could not afford that, indentured servitude.[53]

Recorded accounts of marronage come through the colonial and plantation state, court minutes, letters, jail notices, and runaway slave advertisements: framed as the outlawed, the insurgent, the unruly, and the runaway who steals from the plantation.[54] This archival perspective defines marronage in the context of futility and illegality rather than Black resistance.[55] Maroons found refuge with Indigenous peoples, some of whom had themselves been enslaved. However, relatively little is known directly about North American marronage and independent Native American groups, who eluded observers and left few written records.[56] Histories of maroon activities in North America face "archival silences." Maroons in North

America did not engage with military forces as they did in Latin America and the Caribbean, writes Nevius. "It logically follows, then, that North American maroons did not pen voluminous accounts to leave evidence of hideaways' exact locations."[57]

Maroons were opportunistically leveraged by authorities to tame what was otherwise considered wild. Such representations of unruly maroons intersected with the forbidding swamp itself. The lower swamps of the Mississippi River Delta also provided protection for the Houma and Chitimacha and other first peoples during pressures of settler colonialism. Yet it should also not be lost that the Barataria Swamps served as a protective enclave for piracy and illegal smuggling. The French privateer Jean Lafitte was a prodigious smuggler of West Indian African slaves after the United States banned imported slaves in 1808.

MISCEGENATION AND ERASURE

The intersection of swamps and their inhabitants not only provoked concerns of ambush and resistance to colonial authorities; they also represented a challenge to racial hierarchy that was established and crystallized by the Enlightenment. With the ascendence of the idea of the sovereign human, the Enlightenment also gave us the concept of the *subhuman*—races "trapped" in timeless cycles of nature—as opposed to the Western white rational thinker on a teleological march of progress. Thomas Chatterton Williams says racial identification functions as a veritable prisoner's dilemma. "The idea of distinct human races, as we understand it today, only stretches back to Enlightenment Europe, which is to say to the 18th century," he writes. "I have stayed in inns in Germany and eaten at taverns in Spain that have been continuously operating longer than this calamitous thought."[58] Nowhere was race more important than in the New World, where racial definitions "emerged from a fundamental imbalance in power among social groups." On slave ships transporting men, women, and children to the New World, European captors became white, and their African captives became Black.[59] The establishment of a white European identity in the New World required the existence of subhuman racialized categories. This justified plantations' practice of using enslaved Africans, a practice that intensified after the ban on imported slaves in 1808 and the rise of domestic slavery.

With New Orleans's emergence as the slave capital of the New World in the nineteenth century, domestic slavery required the reproduction of enslaved labor, which turned plantations into breeding grounds through the sexual assault by plantation owners of enslaved women and the separation of families. As Smith writes, "Sexual violence was ubiquitous throughout slavery, and it followed enslaved women wherever they went."[60] The enslaved individual was not only up against the physical power of the assailant, but the power of the state, the power of patriarchy, and the power of society. "These acts were not only permissible but legally encouraged."[61] The coup de grâce was that racial designation was dependent on the mother, which allowed plantation masters to sire as many off-spring as possible—thereby increasing the number of enslaved people and the masters' wealth rather than diluting it by producing legal heirs. This required that mixed-race people—whose skin tone was lightened by miscegenation between owner and property—be legally classified as nonwhite chattel. "The social and political, as opposed to scientific, significance of the binary is obvious in the mad-deningly whimsical nature of one colonial law, which first declared the legal—and therefore racial—status of mixed-race children to be transmitted via the father, only to be subsequently reversed to the mother." This potential liability becomes profitable where the mixed-race offspring are a source of more wealth instead of a drain on it.[62] The children could then be sold off. It is estimated that about one mil-lion enslaved people were separated from their families.[63] Smith says, "In *Soul by Soul*, the historian Walter Johnson writes, 'Of the two-thirds of a million interstate sales made by traders in the decades before the Civil War, 25 percent involved the destruction of a first marriage and 50 percent destroyed a nuclear family—many of these separating children under the age of thirteen from their parents.'"[64]

There is currently only one plantation in Louisiana that narrates the antebellum period from the point of view of the enslaved. The narrative position is stunning. The owner of the Whitney Plantation, a white southern lawyer, John Cummings, decided to dedicate an archive that interrogates the pastoral antebellum luxury that River Road tourism has been known for promoting. Cummings reports to the journalist Clint Smith that his research revealed the banal brutality of it all. In oral histories of former slaves conducted by the WPA, Cummings said he found one account after another of forcible rape and quotidian brutality wreaked on regular people. "I kept looking for an account that did not involve it," Cummings said. "But I never found one."[65] Writing in 1897, W. E. B. Du Bois argued that the science of race did not add up: "When we thus come to inquire into the essential differ-ence of races, we find it hard to come at once to any definite conclusion. Physical characteristics are inconsistent. Color does not match texture of hair, nor size of head, nor tone of skin. Unfortunately for scientists, however, these criteria of race are most exasperatingly intermingled." The differences of men, he wrote, does not explain all the differences of their history.[66]

THE SLIPPERY CLASSIFICATION

One is racially classified differently by different laws, customs, and countries. The term "black" as it was used by British in the nineteenth century applied to anyone from Africa, the West Indies, India, Pakistan, Bangladesh, Sri Lanka, and even Latin America.[67] The one-drop rule in America said that with a drop of nonwhite blood, one was considered *not white*; in Brazil, a drop of white blood classified one as *not Black*. Irish, Italians, and Jews were all, at one time during their American immigration, considered to be colored. "In color theory there is no such thing as

White—it exists solely in our perception of the world, not as a color per se but as the absence of such," writes Chaterton Williams. In real life, too, the lived experience of 'Whiteness' is often construed as the absence of racial identity."[68] It is the neutral point from which all else deviates, which is a move to reinforce or justify uneven power relations. Legal pressures to racially categorize Native Americans, for example, as either "Native" or "Black" nearly wiped them from the historical record in many places. In Louisiana, if one was of mixed race, one was termed a "Mulatto," a slippery classification created in the eighteenth century that gained so much currency that by the twentieth century it could be applied to anyone with a portion of nonwhite lineage.

Chatterton Williams, who is biracial and self-identified as Black growing up in Texas, writes about his own racial dysphoria after siring a child in France with blond hair and blue eyes. His own father, who is African American with a light complexion, said the child was "high yellow," which is one of many terms baked into the American racial psyche. Such terms are familiar to anyone from New Orleans—which was more inventive than most American cities in racially classifying people based on skin tone, eye color, and cultural and linguistic heritage. Among such striations were Octoroon, Quadroon, Creole, Mulatto, and Red. In New Orleans, free people of color sometimes owned enslaved people.

Racial categories—before and after Emancipation—also wedged internal differences between Black Americans and Amerindian peoples. During Jim Crow, the "one-drop rule" that marked nonwhite blood as "colored" forced Indigenous people to identify as white or Black, further erasing Indigenous identity between 1920 and 1964. Intermarrying or partnering by Native Amerindians with nonwhites largely erased their indigeneity in the eyes of the state. Therefore, knowledge of nonwhite people when it consisted of African, African American, and Amerindian lineage was lost within the structures of white supremacy. Indigenous families likely also subdivided among themselves to differentiate between those who—based on outside association of race to location—may have disassociated themselves from other Indigenous families who were identified as having mixed heritage. European chroniclers left a trail of terms such as "mestizo," "zambo," "metis," and "half-breed" to describe individuals who had either African or Amerindian parentage.[69] "Many part-American, part-African persons (with no European ancestry) could easily be subsumed under a racial term applicable to 'pure-blood' Africans, and would not in any case be especially recognizable to most observers as being part Native American," writes Jack Forbes.[70]

The archival record itself was narrated by white authorities, which further complicated racial and Indigenous tribal tension that was often exploited by colonial powers. By 1915, census takers and lawmakers did not distinguish indigeneity among nonwhite groups, eventually conflating the many lineages of nonwhite Native Americans as either Creole or Mulatto. The law did not distinguish between African Americans and Creoles with Indian lineage, which "has implications for the study of the diffusion of cultural traits in areas as diverse as folktales, music, social structure, folk language, and religion."[71] Cultural adaptation by Africans, Caribbeans, Indigenous, and Europeans created the "creolized" culture of New Orleans food, festivals, music, and street culture. In one unique New Orleans tradition, for example, the Mardi Gras Indians, also known as black masking Indians, performatively honor the memory of the city's earliest inhabitants every year with hand-sewn and intricately beaded outfits representing different "tribes," or neighborhoods, of the city.[72]

The Mardi Gras Indians also recount collaboration with Native Americans as well as Black Americans who identify as having Native American ancestry. The performances themselves are oral traditions that operate independently of official archives. "Constructive African and Native American exchanges and merging traditions are absent from most standard narratives about colonialism in the Americas. African and Native American peoples have met in a diverse set of circumstances during the last several centuries." On battlefields they were sometimes allies and sometimes enemies. "Both endured the harsh work and punishment of forced labor systems, but when opportunities for rebellion or escape arose, they formed alliances."[73] The identity of "Black Indians," as the Native scholar William Katz called the people of the African Diaspora who intermarried or were formerly enslaved by Indigenous Nations, is still a relatively unarticulated, and thorny, cultural lineage.[74] Black citizens of the Muskogee Creek Nation, once known as Creek Freedman, continue to struggle for federal recognition and benefits that federally recognized members of the Creek Nation receive. "They were among the thousands of African Americans who were once enslaved by tribal members in the South and who migrated to Oklahoma when the tribes were forced off their homelands and marched west in the 1830s."[75] Freedmen descended from emancipated slaves who were owned or intermarried with the Five Tribes—Cherokee, Choctaw, Chickasaw, Muscogee (Creek), and Seminole—whose societies included enslaved and freed African Americans. Some of the descendants have won lawsuits seeking inclusion in the Cherokee Nation.[76]

The existence of a large group of "Red-Black People," who are part Amerindian and part African, has largely been overlooked until recently. They were usually classified as "Black" by scholars and legal statutes. Some notable figures in this group are Paul Robeson, Josephine Baker, Bunk Johnson, Lena Horne, Pearl Bailey, Leadbelly, and Tina Turner, who in memoirs, biographies, and autobiographies make specific reference to Native American ancestry. Jack Forbes, writing in 1984, using somewhat archaic language, nonetheless describes today's African American population as a composite of African, white, and Amerindian elements: "Considerable controversy has developed at times among white writers as to how much American ancestry Afroamericans actually possess. This controversy is apparently not found among Afroamericans, many of whom have told this writer of their Indian ancestry," Forbes explains.[77] Many Seminole Indians and Black Seminole descendants likewise do not share strong, joint cultural or political activity, despite centuries of close interaction. Those identified as "Black Seminole" have been defined by various titles: "Afro-Seminole," "Freedmen," "Negro Seminole." Their ancestors were African maroons who found freedom on the Florida frontier in alliance with Seminole Indians during the eighteenth century.

When the influential, and controversial, anthropologist John Swinton performed his limited fieldwork in Louisiana, he failed to identify Indigenous people with African descent. "In the 20th century, John Swinton—the world's worst anthropologist—said in 1911 there were few Atakapas," remarked Jeffery Darensbourg, a member of the Atakapa-Ishak Nation, speaking at the Tulane Gulf South Indigenous Studies Symposium in March 2022. "I happen to know there were several thousand left. What kept him from identifying them as us, is they all had African ancestry."[78] This amounts to dispossession by refusal of recognition, which hinges on a lack of research, and decontextualizes contemporary inequalities and efforts for justice from historical dispossession.[79]

Another Indigenous population began to develop in southwestern Louisiana in the mid-nineteenth century among people who had emigrated from the Carolinas and Georgia. They sought areas that were Indian or mixed-Indian and Black and white families. They came to be identified, pejoratively, as "Red Bone," which comes from the West Indies, where "Red Ibo" (pronounced Reddy Bone) was a label for mixtures of races and was likely pronounced Red Bone in Louisiana and the Carolinas. Dispersed groups of Indigenous people in Louisiana, including the Biloxi, Choctaw, and Pacana, were sometimes called "Seminole" in error and included those of white and African ancestry and culture. "Among the Houma, referred to by local whites as the Sabine from the Spanish word for cypress tree, or 'red and white spotted,' French admixture was common. Like the Red Bones, they were suspected of absorbing blacks and once were rigidly segregated by white power structure."[80] This impetus to delineate whiteness from everything else is part of a modern impetus to control or "purify" the seepage of nature from contaminating authority.

The scholar Bruno Latour has argued that one of the driving impulses of modern man has been a persistent aspiration toward "purification," despite our lived experience of hybridization. In other words, he says, we have never been modern because such purification is impossible to achieve despite our best efforts.[81] "Purification involves the clean construction of a nature (and science) separated off from society and the self, while hybridization involves mixtures of nature and culture." The resulting myth is that the realms of the real, the discursive, and the social are believed to be separate. "That's what moderns pretend to do at least, though in practice they produce all sorts of nature-culture hybrids."[82] We see these impulses not only in the classification of and separation between human and non-human worlds but also within these subworlds: the separation of water from land, the dead from the living, and Black and brown bodies from white ones. In coastal Louisiana, people might be racialized based on which bayou they lived on. "It should be noted also that from at least 1741, the term mulatto is used to designate a certain type of land or soil, sometimes described as a 'black mould and red earth' (1789) or 'the red and mulatto lands' (1883)," writes Forbes.[83] The land becomes imbued with characteristics of the people associated with it, and vice versa.

MIASMIC THREATS

Discourses that disparaged and rationalized draining and conquering swamps that offered protection to maroons and self-liberated enslaved people were also driven by a companion movement within the Enlightenment: the mid-nineteenth century's Sanitary Movement. It postulated that good drainage promoted upstanding morals. The Sanitary Movement rationalized the improvement of public health through the management of space and principles of economic circulation. Scholars point to the movement's emergence with Chadwick's report, *Sanitary Condition of the Labouring Population of Great Britain*, in 1842. The industries of medicine and agriculture stood the most to gain from the Sanitary Movement as fallow and otherwise unregulated lands came under enclosure and surveillance by the state. Chadwick, London's main demographer, blamed typhus as well as cholera on miasmic fumes rising from unenclosed fens or marshes.[84] "Miasmas" were considered vaporous swamp fumes that were assumed to be endemic to place rather than spread by organism. Such fumes were attributed to having both gas and liquid—air plus water. Wetlands were believed to be laden with infectious air that emanated from decaying matter.[85] Chadwick was a former literary assistant to Jeremy Bentham, who is known as the progenitor of the panopticon theory that so affected Foucault. Bentham was an advocate of the upright morals of spatial circulation and discipline. In the nineteenth century, Foucault argues, the state began regulating the well-being of its citizens through statistical demographics and the management of space. This "conduct of conduct" is the foundation for modern government. "A good street is one in which there is, of course, a circulation of what are called miasmas, and so diseases, and the street will have to be managed."[86] This space or milieu was a tableau of uncertainty and possibility. "The milieu needs to be managed because overcrowding or congestion leads to poor circulation which leads to increased miasmas and disease."[87] Circulation was paramount.

New Orleans stood at the nexus of this problematic relationship in the eighteenth and nineteenth centuries. Throughout most of the nineteenth century, drainage projects were ad hoc and privately funded. The city's average rainfall of 60 inches a year ended up turning these private canals into "beds of garbage and excrement, fit only to generate fever and breed mosquitoes," according to an 1880 Louisiana Board of Health report.[88] Fires were particularly dangerous due to the inability of fire protection teams to navigate effectively through the mud. Victims of fire include the original church sited at the St. Louis Cathedral and most of the early French colonial structures of the French Quarter. An ordinance was passed in 1788 forbidding the buildings financed by the "King's loan" to be constructed with cypress wood.[89] As the city grew, the complexity of levee and drainage designs increased. Levees spread up and down the river and its connecting bayous. Eventually, authorities eyed the back-of-town swamps. They called for a functional drainage system to enable more people to live in "reclaimed" areas that were once uninhabitable and consequently more exposed to infrastructural disruptions during major storms and floods. Before 1835, the city had invested nearly $5 million in streets, drains, and elevated banquettes. But gutters and canals clogged with subsurface seepage from backyard privies and mud, which made cleaning and clearing them a never-ending task.[90] In 1835, the city awarded the New Orleans Drainage Company a twenty-year charter to drain the cypress swamps between the riverbank and Lake Pontchartrain. Between 1833 and 1878, more than 35 miles of drainage canals were dug across the natural levee back slope and through the lower-lying swamps.[91]

Diseases were thought to be endemic to the surrounding swamps of New Orleans, so it was important for the city to "domesticate" its landscapes to protect itself from its surroundings.[92] Such measures included not only drainage, but shooting cannons to dispel vaporous air. Officials also speculated that contagion of diseases spread from the cemeteries. The city council carried on a prolonged controversy with the wardens of the Cathedral to move St. Louis Cemetery to some other location as cholera, malaria, dengue, and yellow fever claimed the lives of thousands of citizens.[93] Yellow fever was by far the deadliest. Over 100,000 Louisianans, including nearly 40,000 New Orleanians, died from yellow fever between 1796 and 1905. The worst outbreaks occurred during the late 1840s to late 1850s when at least 22,500 residents perished. The so-called Yellow Jack seemed to be a chronic, albeit cyclical, part of life in the city. It tended to arrive in the summer and fall and dissipate as the months cooled. It was a visible and horrible disease. In mild cases, infected persons would feel muscular pain, probably vomit for several days, and swing from intense chills to intense fevers. In severe cases, the skin would turn yellow as the disease incapacitated the liver, kidneys, and heart. The infected victims would then vomit digested blood that had turned black. New Orleans experienced twelve yellow fever epidemics in thirty-five years.[94] As late as 1887, with rival cities such as Memphis, Tennessee, embracing sanitary reform, Charles Dudley Warner visited New Orleans on assignment for *Harper's*. He was stunned by "gutters green with slime[,] . . . canals in which the cat became the companion of the crawfish, and the vegetable in decay sought in vain a current to oblivion."[95] Perennial outbreaks threatened not only residents but also financial development and investment in a port city that counted on regular visits of people and shipments of goods. Foreign businesses often shunned New Orleans as too great a health risk for commercial investment.

The dreaded "late-summer plague" forced public quarantines of riverboats. Costs of disrupted trade were continually weighed against the social cost of

outbreaks. Some outbreaks were exacerbated by the suppression of public information. There are numerous cases of yellow fever epidemics that for weeks went
unreported by newspapers and authorities concerned about hurting business at
the docks. This led to more deaths of workers, visitors, and city residents.[96] Those
with resources fled north across Lake Pontchartrain to the piney woods of what
is now Mandeville or east to the Mississippi Gulf Coast. As a result, poor whites
and African Americans bore the brunt of the scourges. As is true today, the poor
of New Orleans suffered more than the rich because of inequitable residential geographies, where the poorer sections of town lie in the lower topographies and lack
resources to evacuate.[97]

The longest and most fatal US yellow fever outbreak started in New Orleans
with the arrival of an infected sailor in May 1878. Some historians have speculated
that it arrived from Havana. It continued to take victims through June and July
and travel from city to city along the disease vector of the Mississippi River. New
Orleans lost 4,600 lives. Memphis lost 5,000, which was 20 percent of its 40,000
residents. By December, yellow fever had struck parts of Mississippi, Louisiana,
Tennessee, Kentucky, Georgia, Ohio, and Missouri. After traveling from New
York City to New Orleans in September 1878, one dry goods merchant expressed
astonishment at the extensive reach of the fever: "The country between Louisville,
Kentucky and New Orleans is one entire scene of desolation and woe."[98] Amid the
terror, residents of Jackson, Tennessee, placed armed guards on incoming roads
to turn away anyone attempting to enter. Towns in Texas refused trains, mail, and
people from New Orleans for fear that they would be infected. "Shot-gun quarantines," the editor of the *Memphis Appeal* reported later, "were by this time (the 26th
of August) established at nearly all points" in the Mississippi Valley. The *Washington Post* noted "a first-class panic in . . . small towns and villages" surrounding
New Orleans.[99]

Seeking to quell public hysteria, a consortium of New Orleans physicians in
1878 issued a public treatise stating that yellow fever was a specific disease "that
had once been exotic but was now domesticated or endemic." Since quarantine had
never prevented the occurrence of either isolated cases or epidemics, the physicians protested it. They blamed yellow fever on the city's unsanitary conditions but
could not explain why the pestilence prevailed only in the summer. They issued
calls for urban improvement and a comprehensive program of sanitary reform in
keeping with late nineteenth-century discourses of the reform movement. They
called for paving and cleaning of city streets together with efficient disposal of
garbage in the Mississippi River. And they called for a safe and adequate municipal
water supply.[100] "As absolutely necessary" preventive measures, the New Orleans
Medical and Surgical Association recommended proper drainage of the city,
including an underground sewer system and abolition of the backyard privy, or
outhouse. The physicians' report calculated that residents deposited over two million pounds of human excreta into the soil annually, which was "the most difficult

problem connected with the sanitation of New Orleans." New Orleans had forty-four thousand privies. Of these, inspectors declared that over half were "foul" or "defective." These devices introduced sewage into an already saturated ground.[101] The physicians outlined a comprehensive program of sanitary reform.

Ironically, it was an aversion to mud that may have aided the spread of yellow fever in New Orleans. To avoid drinking muddy river water, residents relied on backyard cisterns, which were breeding grounds for the chief disease vector, the *Aedes aegypti* mosquito (which also carries the zika virus). The mosquito breeds almost exclusively in and around houses—in containers such as drinking cisterns, tanks, buckets, roof gutters, and bottles filled with rainwater. It also breeds in flower vases and icebox drainage pans. While the Pasteur Institute had by 1880 exonerated vaporous swamp fumes, or "bad air," a term that stems from the Italian *mal'aria*, the miasmic theory persisted for years. It was largely due to the persuasive link between the revulsive sight and smell of fecund spaces with sickness.[102] In the case of yellow fever, a causal viral organism was suspected but not actually isolated until 1928. "It has practically never been found breeding in swamps, rivers, lakes, or other places where malaria mosquitoes usually breed."[103]

After half a century of marginally successful privately financed public works projects, the city embarked on a major improvement program in the 1890s to relieve the soil of its "soggy conditions," writes Craig Colten.[104] The Drainage Commission of New Orleans was formed in 1896 and developed a $27 million drainage plan. "By 1905, workers completed 40 miles of open and underground canals, hundreds of miles of drains and pipes, and six pumps draining 22,000 acres at 5,000 cubic feet per second. The work was not yet half done, but the effects were already apparent."[105] Muddy streets began to dry. Swamp water disappeared. Soils were able to be paved. California-style bungalows started appearing on streets designed for automobiles in areas that were previously marsh. Even the acerbic New Orleans–born author George Washington Cable marveled, "The curtains of swamp forest are totally gone. Their sites are drained dry and covered with miles of gardened homes."[106]

A CENTURY OF LAND RECLAMATION

The inventor largely credited with "conquering the swamp" was Albert Baldwin Wood, a local resident who designed the Wood screw pump, which was shaped like a corkscrew and could pull 10 million cubic feet of water out of the "soup bowl" of New Orleans. Wood's system drained the "floating land" of its excess sub-surface moisture. The famous Baldwin screw pump was patented in 1912—it was still being used by the city when Katrina struck in 2005—moved water through the drainage canals and up and out of the city. The Baldwin pump is credited with expanding New Orleans's urban footprint to its existing scope. More pumps, canals, and levees were built. By 2005, there were twenty-two drainage-pumping stations in New Orleans with the pumping capacity to empty a 10-square-mile

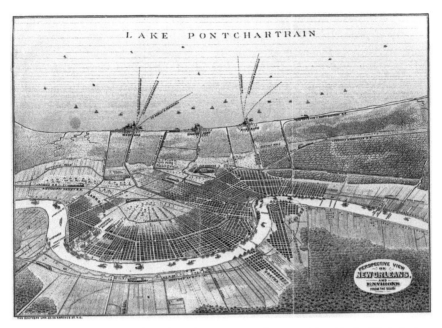

FIGURE 4. Perspective 1884. A modern city emerges on the crescent bend in the Mississippi River as urban development rises on former swampland methodically drained by outfall canals. Image courtesy of the Historic New Orleans Collection, 1974.25.18.125.

lake, 13.5 feet deep, every twenty-four hours.[107] These same canals would later give Katrina's floodwaters access to the heart of the city.

The new sewerage system approved by voters would use "the diluting power of the Mississippi River" and replace old drainage canals with pumps that force effluent through closed pipes up the natural levee and into the river at a discharge point below the city.[108] But not everyone was served equally, of course. The city's Black population typically occupied the swampy back-of-town areas toward Lake Pontchartrain. While the Drainage Commission had undertaken an ambitious Progressive Era citywide drainage program of pumping stations and canals, the coinciding Jim Crow policies challenged principles of social equity by denying services to nonwhite neighborhoods and prohibiting African Americans from leaving them. City ordinances and later deed restrictions legally obstructed deseg-regation. Gaps in the system became apparent in the 1920s. Vast tracts of lakefront property drained after 1920 became entirely new subdivisions, and ordinances and racially restrictive deeds effectively closed them to African Americans. It wasn't until the 1930s that engineering concerns seemed to overcome the prevalent rac-ism of the day when the sewerage system reached previously unserved areas. But this also meant that even when black New Orleanians received drainage and sew-erage services in the 1930s, they were limited to the lowest sections of the city.[109]

Draining the old swamps triggered an early twentieth-century real estate boom that saw a 700 percent increase in the city's urban acreage and an 80 percent increase in assessed valuations during the same period.[110] Most of these lowland lots were not developed until after World War I. Meanwhile, the disappearance of the cypress swamps behind the city through drainage and land reclamation led to tree root die-offs and ground subsidence.[111] In fits and starts, the practice of draining uninhabited swamplands for neighborhood development continued through the late 1980s and not only expanded the city footprint of New Orleans, but opened new areas for development in neighboring parishes.

As the city footprint expanded to the edges of Lake Pontchartrain to the north and to the wetlands of New Orleans to the east, developers used suction dredges to build levees, allowing them to fill in local sands and clay material from below the water bottom. The dredged materials were piped as slurry over varying distances and discharged at the point of levee construction. But these last developed neighborhoods in the city are today more difficult to maintain. While natural levee ridges are easily protected from both river floods and storm-induced tides, the level of river floods may stand as much as 20 feet above the drained flood basin surfaces. Storm-generated tides may be even higher. Hurricanes Betsy in 1965 and Camille in 1969 inundated large areas of the drained flood basin of New Orleans, providing ample proof of the undesirable nature of developing reclaimed marsh and swampland for urban use. Yet developed they were. As Richard Campanella explains, "Modern drainage thus enabled the crescent-shaped city of the 1800s to expand into the spread-eagle-shaped metropolis it is today."[112] But it came at a cost. The drainage system was so successful in removing water from the soil that it opened air cavities where organic matter decomposes, shrinks, and creates more cavities. Fine sediment particles collect and consolidate. Campanella continues, "Half of greater New Orleans would subside below the level of the sea, into a series of bowls—even as they were paved, further reducing the soil's absorption capacity and increasing runoff. Each paved bowl required that the pumps do more and more lifting of more and more water."[113]

Meanwhile, more than 120 miles of subterranean canals underlaced the city. Pumps located in the interior of the city required that the lifting of water be done at the pumping station well before it reached its discharge point, which raised water levels in the outfall canals rather than at the end of the canal just before being pumped into Lake Pontchartrain. "All that stood between high water and low neighborhoods were thin floodwalls." Pumps originally located behind populated areas were now surrounded by these areas. "Unbeknownst to new residents, their exposure to hazard grew with every centimeter that neighborhoods sank, as did their dependence on pumps and barriers to prevent rainwater or seawater from pouring in."[114] A now-fateful decision in 1895 to expel runoff east into Lake Borgne rather than north into Lake Pontchartrain changed the positioning of the pumps. Had they looked to Lake Pontchartrain, the pumps likely would have been positioned along the lakeshore—which would have added protection against incoming storms. "The pump themselves would have acted as gates."[115] In other words, Katrina's damage throughout the interior of the city of New Orleans may have been avoided.

The built environment of canals, levees, and seawalls created the fiction of a dry city, though it had sunk by 3 meters in some areas. This would force a greater reliance on levees and floodwalls, which perpetuated a devastating cycle of groundwater removal, flooding, and vulnerability. This is compounded by the long-term problem of coastal erosion in South Louisiana. The surrounding salt water of the Gulf of Mexico creeps ever closer to a dense urban population that is living below sea level.[116]

This positive feedback loop is one paradox of Extractive Thinking, which I argue throughout this book stems from a philosophy that traces its origins to modernity itself. "Construction interferes with the land-building process: levees contain the silt needed to replenish the lowlands, dredging loosens the land by killing freshwater plants, floodgates and reservoirs further aggravate marsh subsidence." To abandon these projects invites economic disaster. But to continue as before is to invite a worse catastrophe. The system that offers prosperity and security is also consuming the earth beneath our feet.[117]

SUSTAINABLE DEVELOPMENT DISCOURSE

Local observations toward the last quarter of the twentieth century began to note the disappearance of fields and marshes behind the levees and along the Louisiana coast. Around the same time, new schools of thought emerged that began to change the political landscape regarding natural resources. In 1962, Rachel Carson's *Silent Spring* was published, pointing to the damage inflicted by pesticides on bird and aquatic species in California's Central Valley. And then, after being polluted for decades by industrial waste, an oil slick on the Cuyahoga River at Cleveland caught fire in June 1969. The public spectacle is often cited as the catalyst for the creation of the Environmental Protection Agency in 1970, which began reformulating what had otherwise been thought of as untamed wilderness. Environmentalism coincided with legislative developments that included the federal Clean Water Act, first passed in 1960 and amended (generally in a more stringent direction) five times over the next twenty years; the Endangered Species Act of 1966 (updated in 1969 and again in 1973); and the National Wild and Scenic Rivers Act of 1968, which barred or severely restricted new project development on listed rivers throughout the country. At the same time, an important judicial development was the granting of "legal standing" rights to environmental groups, allowing them to bring suit before courts and administrative agencies on the grounds of the public interest.[118] The Santa Barbara oil spill in 1969 and Earth Day in 1970 cemented what became the modern environmental movement.

In late 1969, Congress passed the Environmental Policy Act, which requires federal agencies, including the Army Corps of Engineers, to assess the environmental damage of proposed projects. The following year, Ralph Nader published his book *Water Wasteland* about the destruction of the Chesapeake Bay habitat, and Congress subsequently rewrote the Water Pollution Control Act. It is hard to refute a turning point in the 1970s toward a new environmental ethic.[119] This new age of environmental sensitivity would also affect Louisiana. In 1971, the state legislature established the Louisiana Advisory Commission on Coastal and Marine Resources, which provided a foundation for the establishment of Louisiana's Coastal Zone Management (CZM) Program in 1978. By the end of the decade, the state of Louisiana officially recognized that its wetlands were eroding.

A new rhetoric transformed swamps and marshes, which for almost three centuries were generally repelled by the urban inhabitants in New Orleans, into something that environmentalists and biologists called "wetlands." Lynn A. Greenwalt, director of the US Fish and Wildlife Service (FWS), issued a comprehensive report in 1977 on the classification of wetlands, officially acknowledging that wetlands and deepwater habitats are essential breeding, rearing, and feeding grounds for many species of fish and wildlife. This report expanded an initial inventory conducted by the USFWS in 1954—which at the time was to assess "valuable waterfowl habitat." That report described twenty wetland types. Greenwalt's report was more comprehensive and has been called "one of the most common and most influential documents used in the continuous battle to preserve a valuable but rapidly diminishing National Resource."[120] The Fish and Wildlife Service adopted the new wetland classification system while acknowledging there is no single, ecologically sound definition for *wetlands*, "primarily because of the diversity of wetlands and because the demarcation between dry and wet environments lies along a continuum." Under the heading, "Wetlands and Deepwater Habitats," the report reads, "Marshes, swamps, and bogs have been well-known terms for centuries, but only relatively recently have attempts been made to group these landscape units under the single term 'wetlands.' This general term has grown out of a need to understand and describe the characteristics and values of all types of land, and to wisely and effectively manage wetland ecosystems."[121]

In his foreword, Greenwalt pointed to other uses of wetlands: "[They] perform important flood protection and pollution control functions. Increasing National and international recognition of these values has intensified the need for reliable information on the status and extent of wetland resources."[122] As such, it appears that wetlands emerged from a government inventory motivated by perceived scarcity and anthropogenic value. Wetlands also became commodified as eco-services for recreational and taxable hunting and fishing that brought tourists and vacationers into forested areas. Arguably along this fracture, Louisiana's swamps and marshlands finally emerged as something other than "wasteland" and therefore worth protecting. But this occurred, not through ethical stewardship, but through a value system that prized their utility in providing a protective buffer for oil and gas infrastructure from storms and for commercial services like tourism, fish hatcheries, and waterfowl flyways.

This representation conjures both mud's and wetlands' role in a complex ecosystem of nonhuman actors, as well as their vulnerability that should be protected for capital extraction. It emerged around the same time as discourses on sustainable development, which governs the context under which wetlands are viewed today. The political ecologist Arturo Escobar tracks sustainable development discourse to a 1987 report of the World Commission on Environment and Development convened by the United Nations under the leadership of Norway's former prime minister, Gro Harlem Bruntland. The impetus was fostered by Club of Rome reports of the 1970s, which provided a distinctive vision of the world as a global system where all parts are interrelated, thus demanding management of planetary proportions. The Club of Rome reports argued that nature can be managed scientifically—much like the scientific management of labor—and thus reframed nature as commodity. Escobar argues that this reframing is an attempt by sustainable development discourse to reconcile two old enemies—economic growth and the preservation of the environment—without any significant adjustments in the market system: "This reconciliation is the result of complex discursive operations of capital, representations of nature, management, and science. In the sustainable development discourse, nature is reinvented as environment so that capital, not nature and culture, may be sustained."[123] As wetlands entered the nomenclature, its discursive function and vulnerability to industrial and commercial threats accompanied it.

WHOSE WETLANDS?

Wetlands coexist discursively with an entire apparatus of value and scarcity that is inimically threatening these same wetlands. What were referred to as swamps in the nineteenth century—with their noxious fumes and miasmas—were replaced by capitalist, neoliberal systems of valuation, which focus on the amount of money the wetland commodity could generate in terms of ecotourist dollars, valuable estuaries for seafood and fish hatcheries, protection of infrastructural pipelines from storms, and habitat for waterfowl migratory flyways that are important to hunters. In Louisiana, the perceived value of coastal wetlands is tied to their value to industry along the coast.[124] Today this discursive stamp runs throughout the Louisiana Master Plan for a Sustainable Coast, as its authors point specifically to the financial importance of wetlands to the area's economy in order to justify investments to protect them: "Experts have tried various ways to put a value on the coast's abundance, more in the spirit of highlighting the incredible gifts of our landscape than out of certainty that these gifts can be perfectly captured in numbers. One of the ways researchers assign value to natural systems is by considering what are known as ecosystem services, meaning the benefits that the environment provides to people. In Louisiana, these benefits range from oyster and shrimp fisheries to flood reduction, to nature-based tourism."[125]

And according to the master plan, these benefits have a dollar value. This reconciliation between nature and capital allows the state to move forward with mitigation plans that place eco-services within the same capital calculus of valuation as all other surplus value commodities. Of course, those that are most lucrative, such as oil and gas infrastructure, stand at the front of the line for coastal restoration protection.

The three-hundred-year effort to separate water from land tends to map onto a project of modernity to not only finish God's second Eden by making "fallow land" productive but also separate humans from nature. Today, contemporary schools of environmental science recognize the efficacy of sustainable practice in one form or another to sustain human communities and/or capitalist systems operating under scarce resources. But we might also ask what politics are foregrounded by positioning marsh and mud as commodities and protectors of cities and infrastructure? How does this arrangement naturalize the infrastructures and cities—and perhaps modernity itself—and frame marshlands and mud as almost a fungible utility? Their associated value lies in how they are used and manipulated, which continues to place them within a constrained value system. An alternative is nearly impossible to imagine if we continue to use the same canvas from which these questions themselves are drawn. To conjure New Orleans or Louisiana risks reproducing a discourse about land that is predicated on extraction.

Interlude
——

Toledo Bend

In second grade, you remember crawling over the muddy banks to reach a rope swing at the edge of Toledo Bend. The swing would cast you clear over the surface, and you could dive or flip into the warm Louisiana water. But you had to endure the oozy texture of muck that caught under your fingernails as you clawed your way up the steep, slick embankment. The minerals and sediment of the water turned your bathing suit to an auburn hue.

Two alligator-size bullfrogs by the dock struggled against a pair of large hands putting them into an Igloo. You had never seen creatures like that before.

You normally wore an earplug for protection. A tube to drain fluid was inserted that year. But you forgot it in its red plastic pouch at home. Dad fashioned you a cotton ball with some kind of wax-like texture on it. The rope swing was pushed in your direction. The twined rope was slippery and gritty with lake sediment. You pulled it back as far as you could, back a step up the slippery bank, and leaped onto a knot in the rope where you scraped your pressed thighs that were vice-gripped around the rope.

You swept over the water in a pendulum arc that hung in an apex just enough for your release into the glassy surface. As you crashed into the dark water, you furiously stroked toward the surface in mid-descent, prying yourself away from the furry bottom.

3

The Villainous River

At the Lower Ninth Ward, three fishermen sit in folding chairs perched among riprap boulders inside the Mississippi River levee slope. A walking path atop the levee is dotted with benches next to overflowing garbage cans. Dull humidity is occasionally teased by a gust of turbid air over the brown surface. Red ant bites from the unkempt grass along the backside slope needle my ankle. I slap at a horsefly on my neck. One of the fishermen fusses with a line hung up on a rock. Downriver, a line of tankers muscling against the current align like jetliners to a landing strip. This location—affectionately known by locals as "World's End"—feels like the backstage of a technical performance. Industry echoes across the ribbony water. A school of three ducks float on a current as if on an ice flow. An amusing thought in this August heat. Every couple of minutes, a bellowing horn communicates a ship's position. Currents in the middle of the river crest like class 2 whitecaps. A small transport boat, *Miss Emerson*, awaits an approaching downstream tanker making its slide around Algiers Point opposite downtown. This turn is treacherous. Pilots liken it to driving an 18-wheeler on ice.[1] In mid-June 2019, a barge hit the shore in Algiers, knocking over a utility pole and causing an outage for five thousand people.[2] As it swings around, the bow of the tanker points toward the opposite bank and its stern points east to the French Quarter. It turns in slow motion as *Miss Emerson* pulls up beside it. Two men climb out of the smaller vessel onto the ladder on the belly of the tanker. They will navigate the remaining 106 river miles to Pilot Town, which is accessible only by water or air, at the Bird's Foot. Local pilots, members of the Crescent River Port Pilots Association, exert considerable political influence in the state. They earn more than $520,000 a year, which is set by the state legislature. Each year, the pilots's association hosts a delegation of lawmakers at Pilot Town for "educational" purposes. For many of the most experienced pilots, the year 2019 was their most difficult on the river. It had been in flood stage for a record seven months.[3]

After the pilot climbs aboard the tanker, it appears from my distant vantage point that *Miss Emerson* almost nudges the tanker into its turn before backing off and readying for the next tanker coming upstream from Pilot Town. Meanwhile, the nineteenth-century *Creole Queen* paddle wheeler makes its sliding turn behind the tanker as it takes its usual round of tourists to the Chalmette Battlefield to recite the Battle of New Orleans.[4] On the swollen river, debris surfs along the currents. The high stage of the river has inundated the normally sandy forests that sprout up in the sandy batture forest inside the levee. By July 2019, the immediate threat of the levee overtopping had passed, but the river level towered over the Ninth Ward rooftops down the slope of the grassy side of the levee. Strange that such a specter could be normalized.

A pair of tugboats are moored against rusted wharves. There is little of what is perceived as the "natural world" in this industrial corridor—even as the men next to me cast their lines in the water. Their buckets are empty, but they had just arrived. "Y'all leave any fish in the lake for us?" ask three newcomers, who stake a place among the concrete just inside the point where the levee forks into the Industrial Canal and cleaves through the Ninth Ward, forming the Upper and Lower Nine. A pair of tugboats with barges wait in the canal for the St. Claude Avenue drawbridge to rise. This geographic peculiarity opened this neighborhood to Katrina's wrath. Water was funneled here from the Mississippi River Gulf Outlet (MRGO), which is a gaping passage that connects the Gulf of Mexico to the Industrial Canal, leaving the sliver of the Lower Ninth Ward of New Orleans and the adjacent towns of St. Bernard Parish completely deluged. The Gulf outlet, known as Mister Go, was closed by the Corps of Engineers after Katrina with a shell ridge down its middle to block commercial traffic, but a fifteen-year stalemate over marsh restoration funding between the Corps and the state prevented reducing its gaping width and leaving vulnerable the hardest-hit communities of the Lower Ninth Ward and St. Bernard Parish that were flooded by Katrina.

The fishermen cast their lines of shrimp and other small bait into the brown water as if it was the most normal thing to do—as if the catch was safe to eat, which they said they would do. I walk down the levee through the tall grass to my car and slowly drive upriver toward the French Quarter. Still, the summer before COVID shutdowns, tourists ply T-shirt shops with Styrofoam cups of daiquiris and overwrought straw hats oblivious to the cat-and-mouse game of steel and hydrology taking place just out of view. More Caribbean than American, New Orleans is widely represented and often misunderstood by its visitors. The Louisiana novelist James Lee Burke once described it as a city where "the air smells of the river, dead beetles in a storm sewer, the wine and beer cups in the gutters, damp soil and night-blooming flowers and lichen on stone."[5] But that doesn't fully capture it.

We keep the river at arm's length. It hides much danger below its surface. It is notoriously dangerous for those who find themselves in the water by accident or on purpose. Large debris—trees, boat wreckage, even fiberglass bathtubs—may move swiftly beneath the surface. It contains vortices and undertows at ever-shifting depths. Even those who wade along the water's edge may disappear, followed by others trying to save them. In April 2022, three teenagers were lost at Algiers Point across from the French Quarter when two of them waded into the water and a third tried to rescue them. The depth there dips to 100 feet, and a permanent eddy scours out a turn as the river swings eastward at over 90 degrees. There are places in the river that will trap a body under water. Calder and Wilkinson quote a river captain, "Even with a life jacket you can be pulled down. You're not going to swim against the current. It's impossible. The weight of the water is so heavy, and the velocity is so strong. You're helpless."[6]

TECHNOSCIENCE AND PATH DEPENDENCE

As a cultural and national symbol, the Mississippi inspired awe. It was characterized as both devil and savior in newspapers, speeches, and music. Some of the earliest forms of activism in the United States were organized around what to do about the Mississippi River. In many ways, the management of the Mississippi River provided a common site for organizing Americans of different geographic areas and political stripes around the modern notion of controlling American waters and ridding them of unruly mud. Through the past two centuries, every flood event represented the river's resistance to control and containment; and over time the river was hardened with higher levees and dams. These added constraints predictably caused a feedback loop of even greater pressure on man-made systems and more violent catastrophes. Extractive Thinking was upheld and reinforced by infrastructure and discourses of the river circulating through the halls of government, newspapers conventions, and various technical reports that reflected attempts to not only control the river, but live securely near it. At any time, the Mississippi could slide back into its preternatural state. Maintaining the Mississippi as a commercial waterway required the continual application of oversight and intervention by the US Army Corps of Engineers, the Mississippi River Commission, parish or county "levee boards," and organized petitions to the federal government. The Mississippi River and its tributaries were described by various interests at various times as a national artery and a great inland ocean that connected a divided land.

Christopher Morris calls the Mississippi our "imagined river," frozen in time and place by memory and earthen walls. Its continued existence relies not only on its cultural importance, but the militaristic determination by the Army Corps of Engineers, which maintains the channel and southern passes by means of a twenty-four-hour dredging and ridge reinforcement project.

As an alluvial river, the Mississippi is quite *unstable* from the perspective of a fixed channel.[7]

ENGINEERING CERTAINTY

In terms of engineering, the problem of the Mississippi was threefold. Its mouth was clogged with sandbars; its channel was ever changing from sediment, flood levels, and erosion; and it would often overflow its banks. All these problems are features of natural alluvial processes. They allow the silt and sediment to enrich soils, build land, and produce habitat. The river's shifting behavior affected the political economy of places it bordered. Preferred interventions, whether flood control levees or removal of rapids, that were advocated reflected local financial interests. These contesting interests were couched in discourses of sovereignty, nation building, security, and even religion. And each had a major influence on river science and policy in the nineteenth and twentieth centuries.

Taming the Mississippi River was an acute preoccupation, particularly for those whose fortunes were tied to it. The river appeared impervious to control. Bottom depths were ever changing from moving sandbars. Islands surfaced indiscriminately. Seasonal floods added orders of magnitude to its current and height. Mark Twain described the Mississippi as a particularly unruly object in his memoir, *Life on the Mississippi*: "A cut-off plays havoc with boundary lines and jurisdictions: for instance, a man is living in the State of Mississippi today, a cut-off occurs tonight, and tomorrow the man finds himself and his land over on the other side of the river, within the boundaries, and subject to the laws of the State of Louisiana! Such a thing, happening in the upper river in the old times, could have transferred a slave from Missouri to Illinois and made a free man of him."[8]

The Mississippi's most famous traveler cut his teeth as a steamboat pilot in the heyday of riverboat culture. In his memoirs, Twain reflects on the constant vigilance required of an apprentice pilot who plied the Mississippi before the age of dams and jetties. The turbid rush of sediment and silt built and washed away subsurface sandbars and protruding islands in a manner so fickle that nineteenth-century riverboat pilots were required to memorize every changing inch, curve, and depth with each new trip, "as if one were dropped at random on the longest street in New York in the middle of an 'inky black night' and must describe every lamppost and doorway."[9] By the return journey, the report had again changed. In fall and spring, the water could rise with "terrifying rapidity and overflow its banks in certain reaches till it is 60 miles wide."[10] The Mississippi was the only river in the world with "mud lumps," or volcano-like geysers of gases and liquid mud. They ranged in height from 3 to 10 feet and could spontaneously lift a passing ship.[11] Twain describes the river of his youth as one whose alluvial banks cave and change constantly. Where underwater trees, called snags, were always "hunting up new quarters, whose sandbars are never at rest, whose channels are forever dodging and shirking, and whose obstructions must be confronted in all nights and all weathers without the aid of a single light-house or a single buoy; for there is neither light nor buoy to be found anywhere in all this three or four thousand miles of villainous river."[12] As a proxy for one of the many secondary sources Twain used to fill out his memoirs, he quotes a traveler, Mrs. Trollope, who described in 1827 approaching a muddy mass of waters that mingled with the deep blue of the Mexican Gulf. "I never beheld a scene so utterly desolate as this entrance of the Mississippi," she writes. "Had Dante seen it, he might have drawn images of another Borgia from its horrors. One object rears itself above the eddying waters; this is the mast of a vessel long since wrecked in attempting to cross the bar, and it still stands, a dismal witness of the destruction that has been, and a boding prophet of that which is to come."[13] Such ghost wreckages were common signposts of caution to travelers, who had few travel alternatives. Rail was not yet common. Horse-drawn carriages invited their own danger. But the river generated a unique set of hazards. Boats that went aground on a sandbar might be stuck for days or weeks until either help arrived or the river rose. Sandbars in the low summer season made the river impassable. Then there were floating trees: saplings, hardwoods, and "ancient giants." They smashed into bottlenecks on tight bends and formed "boxing, clunking plateaus in the shallows." Clumps of them would fasten together through mud and debris. "These were known as wooden islands, and they would go careening down the river for hundreds of miles at a stretch. The unmistakable cracking and grinding echoing along the surface, colliding with whatever was in its path along the shore." Boats that could not be maneuvered out of the way would be pummeled by splintered logs.[14]

Hunting subsurface snags required a particular skill for reading the current and decoding patterns. As Sandlin explains, "A trailing braid in smooth water was a sure sign of a snag; a quilted ripple was a tangle of submerged logs; a line or fold across the water was an undertow; a persistent swirl of froth was a whirlpool, where a strong tributary flowing quickly into the main current had created a vortex beneath the surface. The voyageurs had to teach themselves all these clues by experience and the river put a premium on fast learners."[15]

Snags were so varied that they generated their own vernacular. A tree standing straight up on the river bottom with branches just under the water line was called a "Planter." A "Sleeper" would stick sideways into a riverbank or sandbar stretching out at full length under the water. "Sawyers" waved back and forth in the current in a sawing motion. A "Preacher" bobbed up and down, rising out of the water and falling back like one administering a baptism. Snags could rip a hull or capsize a boat. They were everywhere and invisible to the layperson's eye. "It was traversing a flat and infinitely malleable surface of mud, silt, and clay—and this meant that it was free to move however and whenever its currents shifted."[16] Uncertainty was the rule: "Every day, somewhere along the river, huge bluffs were collapsing; overgrown banks were falling in on themselves; ancient stands of trees were sliding down into the tide. Sandbars were growing into islets. Islets were accumulating rocks, rotten logs, and mud, and sprouting with countless scattered seeds. With every voyage, the familiar was reshaped or erased while new hazards appeared out of nowhere." There was simply no way to confidently map the changing course of the river.[17]

During frequents stops at a critical pilot station in St. Louis, chance may have steered Twain's path across that of James Eads, who clerked at the station. Had the two contemporaries sat for a drink, the conversation would likely have turned on worldviews looking in opposite directions. While Twain's later fame hinged on works that celebrated and lamented the disappearing wildness of a river and bygone frontier, Eads built his fortune causing Twain's lament: taming the river. Like Twain, Eads was self-educated and ambitious. Both men learned their trade from the pilots and practice on the river itself: Twain by piloting the treacherous surfaces, Eads by walking the bottom using salvaging tools and a self-designed diving bell.

PROFITABLE HAZARD

The moving, unstable bed of the river was a treasure trove for adventurous entre-preneurs willing to brave the depths to excavate. Steamboats were lost every week to the violent maelstrom of the changing river. Boiler explosions were common and exacerbated by sediment-laden river water used as coolant in overheated pipes. Hulls were ripped open by underwater tree snags.[18] Valuable equipment and cargo were quickly buried in mud and quicksand. Insurance companies were willing to give salvage wreckers as much as half of the rescued cargoes. Once a vessel or freight had been wrecked for five years, it belonged to whomever could retrieve it.[19] Eads was among the first people to walk the riverbed. At the age of twenty-two, he formed a partnership with a boat builder to begin salvaging the underwater graveyard. He and his team used modified diving bells to descend into the violent darkness. He designed the bells from wooden barrels connected by oxygen tubes to a floating wrecker. Eads described wading through a "snowstorm" of sand and finding little purchase for his feet as he pushed against the constant force that few attempted or understood.[20] Eads would search into the depths with little certainty of what he would find.[21] "Eads and his partners worked up and down the river for hundreds of miles."[22] He walked the bed of the river four hours a day, every day. As Twain published Life on the Mississippi, which looked back with nostalgia on a fron-tier that was disappearing from the national imagination, Eads was busy attempting to eradicate the idea of wildness at its very essence. In an 1878 report to Congress, Eads described the Mississippi River as a force that was both immense and tamable.

> Every atom that moves onward in the river . . . is controlled by laws as fixed and certain as those which direct the majestic march of the heavenly spheres. Every phe-nomenon and eccentricity of the river, its scouring and depositing action, its caving banks, the formation of the bars at its mouth, the effect of waves and tides of the sea upon its currents and deposits, are controlled by laws as immutable as the Creator,

and the engineer needs only to be assured that he does not ignore the existence of any of these laws, to feel positively certain of the result he aims at.[23]

The more capricious the river seemed in its "terrible games," which to others appeared to be the result of chance, the more men like Eads viewed it as an invitation to demonstrate man's control over nature.[24] The nineteenth century saw an almost religious belief in science that would reveal the laws of the river's mechanics and put it to work for men. This modern approach essentially meant erasing and hardening its murky boundaries so that water rather than mud would be its main currency.

In less than a century, the proverbial mudscape would be transformed. Its edges would be leveed, and most of its adjacent alluvial forests would be drained and dried for plantation agriculture and urban settlement. By the time Twain had penned *Huckleberry Finn*, the river represented in the book lived only in the author's memory. Rail had supplanted river commerce due to its safety and speed. The marauding river flotillas and town ports were supplanted by depots and new centers of commerce. Traffic on the river was relegated to barges and bulk traffic. In *Life on the Mississippi*, Twain makes note of the unusual sight of two steamboats at New Madrid, Missouri, just south of the Ohio River–Mississippi River confluence where the Lower Mississippi begins: "Two steamboats in sight at once! an infrequent spectacle now in the lonesome Mississippi. The loneliness of this solemn, stupendous flood is impressive—and depressing!"[25] Steamers by then would be novel and notable for their extravagance, an homage to a bygone era. As *Huckleberry Finn* was enshrining the Mississippi with its iconic status in 1884, the river and its culture had already indelibly changed. According to Sandlin, Twain was waxing nostalgic for a river that its contemporaries in the early nineteenth century decried for its accompanying chaos and corruption. Twain's predecessors saw the river as "crowded, filthy, chaotic and dangerous. . . . Where Twain saw eccentricity and charm, they had seen corruption and unchecked evil. Where he saw freedom, they had seen a jerry-rigged culture swept by strange manias and mysterious plagues, perpetually teetering on the edge of collapse."[26]

PLYING THE WILD

Landings in river towns represented not only the hub of commerce but also danger. Crews were notorious, loud, and boisterous. Drunken brawls spilled out into the backstreets of dock districts. "Few towns were enthusiastic about welcoming the river people. At St. Louis, there was a night watch with 50 armed men assigned to the dock district, just to make sure that the river people didn't stray too far from the levee."[27] A town's river district was often cordoned off. A landing with a "maze of slums and shanties" would be distanced safely away from the town center. Natchez, for instance, was situated high on a bluff overlooking the makeshift shanties built from wood salvaged from wrecked or abandoned flatboats. Disorderly music wafted up from the commotion, along with pops of gunfire and frequent screams. Just before dawn, the procession of boats would break into a "reveille of river horns" and drift out into the expanse of the river.[28] "The immensity of the river, the vagaries of the current, and the crowds of traffic down every bend meant that the next night they'd be sorted into wholly different congregations downriver. It was a rare event for any boats on their way to the delta to encounter each other twice. The river didn't encourage lasting friendships."[29] Flatboats and barges carried everything to the insatiable delta markets downriver: pelts, minerals, soaps, cured meats, lard, coffee beans, apples, pigs, turkeys, horses, whiskey, furs, wheat, rice, and animals. Goods were sold, bartered, or stolen.

River merchants could be found docked at a levee or village to display their goods. Such floating markets of kitchenware or books or tools or even furniture and cabinetry could be found on any stretch of the river. There were gambling boats, smithies, greenhouses, tailors, and fully stocked general stores; there were showboats of traveling troupes; there were doctors and traveling medicine shows; there were musicians and burlesques; and there were mesmerists, homeopaths, and snake oil salesmen. They could pack up and disappear as quickly as they came. Past the great confluence of the Missouri and Ohio, the Mississippi River traffic below St. Louis was described as a jumbled assembly of all types of watercraft: barges, pirogues, keel boats, canoes, schooners, tugboats, ferries, houseboats, and shanty boats—all crammed with cargo and people.[30] Immigrants, itinerant laborers, and enslaved people were carried by so-called soul drivers, which, unlike other river shanty boats, would keep to themselves because they were shunned by the other river people.[31] Often river folks would tie up together for safety. Sometimes families were on the move looking for a better life, carrying with them furniture, children, and farm animals. They anchored for the night, passing on gossip or bartering food or jobs.

As the landscape changed and foliage unfolded into the delta, soughs and bayous and swamps grew around the river. The delta region was described as a partial gloom of Spanish moss and lagoons whose pungent smell was known to sicken travelers. The banks on either side "melted away into an indeterminate ooze" that widened into expansive meadows of reeds and cattails. "There was no firm line between the river delta and salt estuaries. But in the end, the last islets fell away and the great freshwater flood of lime, gold, and brown" emptied into the Gulf of Mexico.[32]

DANGEROUS DUTY

A simple fall overboard could be fatal from the current's turbulence, competing vortices, and undertows, or simply the cold temperature. The main current was hidden within the dark cover of mud and sediment hidden from sunlight. The cold of the murky depths could lead to hypothermia. And the water was so braided with crosscurrents and undertows that a man overboard was generally considered a man lost. "Even if the rest of the crew noticed in time that he was floundering in the water, there usually wasn't much they could do for him; the boats couldn't be turned around against the force of the current, and most were too unwieldy to be maneuvered quickly into shore—assuming the crew was willing to try."[33] A man who did manage to swim to shore might find himself hundreds of miles from the nearest settlement.[34]

Even in the best conditions, danger lurked. Notwithstanding the snags, the river's thick sediment and mud clogged and overheated boilers. An anthology compiled by the Mississippi River Commission details a historic log of spectacular disasters from boiler explosions. The first steamboat had arrived in New Orleans in 1812. By 1830, there were more than 120 there. "Moving upstream at the unheard-of speed of 14 miles an hour, riverboats not only made it easy and affordable to travel and migrate, but they became the standard means for transporting agricultural commodities and manufactured goods."[35]

The very first boiler explosion on the Lower Mississippi occurred in 1817. Just off the west bank of Pointe Coupee, Louisiana, a year-old steamer, named the *Constitution*, was rounding the point when its boiler exploded, wrecking the front part of the cabin and killing or wounding thirty passengers.[36] Black Hawk Point, an infamous location just north of the Mississippi and Atchafalaya confluence, was the scene of two remarkable steamboat accidents. On the first occasion, "the steamer Black Hawk was southbound on December 27, 1837, when she passed this point with a large number of deck passengers, some Army officers, and a shipment of government payroll money on board." The boat's steam boilers blew. The explosion swept some of the passengers and "a number of the boxes of government money" into the river. "The steamer then caught fire and drifted downstream, leaking furiously and burning very rapidly."[37] An estimated thirty people drowned.

Almost two decades later, in March 1854, the *John L. Avery*, loaded heavily with passengers and freight, gathered at the same Pointe Coupee, hit a snag heading upriver at Black Hawk Point and sank. Hogsheads of sugar were stacked along the outside edge of the deck, effectively hemming in all the deck passengers. The boat went down too rapidly for escape. An estimated eighty to ninety people drowned. Most of them were Irish immigrants.[38] Outside of St. Maurice, Louisiana, at mile 271, the colossal *J. M. White*, once called "the finest river boat in the world," caught fire on December 13, 1886, and sank.[39] The 320-foot-long riverboat had been commissioned in 1878, the year after a yellow fever epidemic. Its partially buried wreckage could be seen for years after in low water periods until it was eventually buried in a sandbar. There are an untold number of buried vessels in the alluvial plains adjacent to the river.

One of the most egregious river disasters happened in 1837 to a Creek tribe that was being forcibly relocated by the federal government. Their Mississippi River steamboat, the *Monmouth*, collided with a downstream vessel just north of Baton Rouge. The *Monmouth* was chartered to take the Creek to the Arkansas River for their eventual resettlement in Oklahoma. "Northbound at the island with a heavy load of passengers, the *Monmouth* collided with a sailing vessel called the *Trenton*, which was being towed downstream by the steamer *Warren*." The collision at Profit Island caused the *Monmouth* to break apart. Hundreds were lost in the water. Accounts of the accident were vague and very brief. An estimated three hundred to four hundred Creeks drowned. The toll may have been higher. According to the account by the Mississippi River Commission, "No one really seemed to know or care."[40]

When an accident like that occurred, the smashed property and drowning people were swept downriver. A farmer hoeing corn on the riverbank or a resident walking about in a nearby town might be alerted to the sight of fire in the distance or the haunting percussion of pleas for help that grew louder as the wreckage approached. Visible signs of the accident would be strewn about, with survivors hanging on to shoreline limbs or large flotsam. If the accident happened at night, there would be no point in attempting rescue. "Every traveler on the river got to know the sight of bodies drifting with the current, or hanging from a floating island, or boating among the logs piled up on a river bend—the red shirt that the voyageurs wore, the closest thing the river had to a uniform, could be spotted a mile off like a distress signal."[41] The largest accident occurred on the *Sultana*, caused by a massive boiler explosion whose concussive blast set off the other two boilers, killing most of the crew and all the cabin passengers. The wreckage filled the river with scalded victims, corpses, and floating debris. Boat whistles and church bells rang out from the waterfront as the wreckage passed. A pillar of fire and smoke rose above the hills with the early light of dawn.[42] The bones of such wreckages were made visible by an extended drought in fall 2022—recounted in Facebook posts and newspaper stories of curious residents descending the inside of the levees and batture for the novel walk.

TAMING THE WILD

According to the technology historian Edward Layton, the scientific and technological communities in nineteenth-century America went through a scientific revolution from a "craft affair" similar to that of the Middle Ages, where oral traditions passed from master to apprentice. The new technologist was apprenticed via a college education, a professional organization, and technical literature that was modeled after those of science. By the end of the 1800s, technological problems were treated as scientific ones, and traditional methods and empiricism were supplemented by tools borrowed from science. Changes were taking place throughout the physical sciences, from the engineering branches to chemistry, biology, and geology. "The result might be termed 'the scientific revolution in technology.'"[43] Controlling the Mississippi River would require more than the hubris of oversized egos and the application of grand engineering methods and technologies. It required trial and error, disasters, and responses. Any force applied to the workings of the river would be met with a counterforce in a cunning exemplar of Newton's Third Law.[44] Engagement with the river was predicated on uncertainty, and the ability to control it was critical to the symbolism of the nation's manifest destiny on the continent, as well as the elevation of the Army Corps of Engineers charged with its management. It also required vast appropriations from Congress, which established the Corps of Engineers as the preeminent agent of modern conquest.

Empirical conquest of nature and natives was wrapped up into the rationales for flood control and navigation improvements. Officially created as a war academy and fort-building agency in 1802, the Corps embodied the discourse of the early American period that success of the Union was tied to the development of the vast continent through control and management of its rivers and harbors. The oft-heard phrase of the period was that developing and improving land was to "complete what God had started." President Thomas Jefferson opposed a military elite, but he revived the US Army Corps of Engineers and signed bills to have the Corps build piers, harbors, and lighthouses for civil purposes.[45] The early Americans understood the strategic importance of water communication, which married the nation's commercial aspirations to its security and defenses. "Surveying and science converged in a literature on the strategic importance of water communication," writes Todd Shallat. "Army maps and reports became aids to commerce that marked the defensible limits of territorial expansion."[46] The US Army Corps of Engineers, in exemplifying the two main progenitors of American philosophy—science and government—has been called a "nation-builder."[47]

Shallat has likened the Corps' operating logic to his observations of Max Weber, who marveled at a bureaucracy's ability to affect policy through its vast power of implementation. The Corps, in Shallat's analysis, attempted to acquire jurisdiction over projects through self-promotion and by cultivating local support from regional directors. It also assisted corporations and promoted its own facility for production.[48] Often the rationale for river intervention was based on the path dependency of protecting a previous intervention. A circular logic emerged early that critics to this day continue to attribute to the Corps' modus operandi. If a waterway was navigable, it was important, and worth defending. And once fortified and stabilized, it was worth protecting, if not improving, to facilitate better defenses. Once a river had been dredged, or "de-snagged," the Corps would continue to maintain the waterway regardless of its cost or relevance. Byzantine congressional appropriation bills that contained everything from a pier in St. Louis to resources to clear the Delaware River breakwater turned waterway spending into a jigsaw puzzle of agreements.[49]

SURVEYS AND STATECRAFT

In *Seeing Like a State*, James Scott demonstrates the power of surveys and maps in the execution of modern statecraft. Citizens must adhere to the grammar of the survey and standardized measurements for legal standing. "The categories used by state agents aren't merely means to make their environment legible," Scott writes; "they are the authoritative tune to which most of the population must dance."[50] Administrative practices produce records of ownership, categories of race and ethnicity, arrest records, political boundaries, and economic plans. Modern statecraft is largely a product of internal colonization on behalf of the state itself.[51] Army surveys became beachheads for modern conquest of the American continent, from disciplining wild rivers and cutting canals to opening the west to railroads—often displacing the continent's native inhabitants in the process. River engineers trafficked in this project of empire. The first comprehensive survey of the Ohio and Mississippi Rivers occurred in 1822. US Army Corps engineers confirmed reports that the Lower Mississippi held thousands of submerged trees that were fatal to riverboats, while the river itself, because of its alluvial nature, constantly tried to change course.[52] There was still no consensus on how best to control floods and improve navigation. Engineers recommended that the government devise new ways to clear the rivers, which included removing submerged trees, implementing dams to narrow the river, removing sandbars, and building levees to hasten navigation and prevent flooding.[53]

GIBBONS, FEDERAL OVERSIGHT, AND POLITICAL LOBBYING

Federalist Party leaders, for example, argued that vigorous, central-state programs of internal improvements would create common interests across the regional sectors. In 1824, the US Supreme Court decision, *Gibbons v. Ogden*, held that the commerce clause of the US Constitution gave the federal government the power to regulate river navigation.[54] The *Gibbons* case allowed Congress to direct the Army Corps of Engineers to make navigational improvements to river channels. The General Survey Act of 1826 followed, which appropriated $75,000 to improve navigation.[55] While federal authority over navigation was codified in the 1824 *Gibbons* decision, flood control was delegated to local landowners and districts, which generally led to uneven construction standards. A weak levee upriver might collapse and spread misery to all. Or sluice cuts through the levee might allow the river to moisten fields during times of drought. Yet different crops had different needs, which led to local conflict, writes Campanella: "Rice planters liked to inundate their fields to kill off weeds, whereas those raising sugar cane feared excess water would cause root rot." Everyone fretted over the integrity of the sluice itself. "If the river got too high, if the sluice gate malfunctioned, or if erosive currents scoured underlying soils, a levee might cave in, which could lead to a crevasse—that is, total levee failure causing destructive flooding."[56] Major storms also brought fears that saboteurs would intentionally sever a levee across the river to release water pressure on their own levee defenses. This particularly impacted slave states in the lower Delta, which incentivized cotton and sugar plantations through programs to drain and "reclaim" former swamplands. Without official federal oversight on flood control, any significant levee aid required political pressure on congressional representatives.

NINETEENTH-CENTURY SCIENCE

In the mid-nineteenth century, there was no pure public policy or clear understanding of the river and its mechanisms, despite myriad surveys and efforts to improve the river for navigation and stabilize it in some way against floods.[57] Until 1837, there were no American books with sections on dams and jetties. Craftsmanship was more ancient than the sciences and relied on apprenticeship and builder-to-builder contact. River engineers faced an array of questions: Why do alluvial rivers like the Mississippi weave back and forth like drunks in an alley? Do meanders result from terrain characteristics or from alluvial processes? How does the sediment or bed-load material moving along a streambed affect sediment deposition? Do bed-load particles leap along the bottom like ballet dancers across a stage, or do they slide along in a layer like maple syrup across a stack of pancakes or roll along like bowling balls?[58]

Three decades into the twentieth century, engineers still knew far more about the structures they placed in streams than about the streams themselves. Variables include sediment load, morphology, discharge, and even location. And even more variables are perpetually coming together to affect river flow, such as velocity, channel width, channel depth, gradient, and bed "roughness" (the resistance of a bed to flow). As Reuss puts it, "The challenge is somewhat analogous to designing suits for a customer who is both demanding in his needs and discontented with his shape, constantly indulging in fad diets and binge eating."[59] Experts must come to a consensus on how to design a dam, revetment, or levee for a constantly changing profile. Reuss again: "The answer is to design within a range of anticipated parameters. Still, neither the tailor nor the river engineer will sanguinely predict success."[60] In river engineering, humility is a necessity.

A river's discharge can vary widely for many reasons, not the least of which is human activity upstream, which modifies the floodplain. River engineers came to recognize that idealized fluid mechanics theories that were French in origin and based on Newtonian physics could not fully account for the many forces

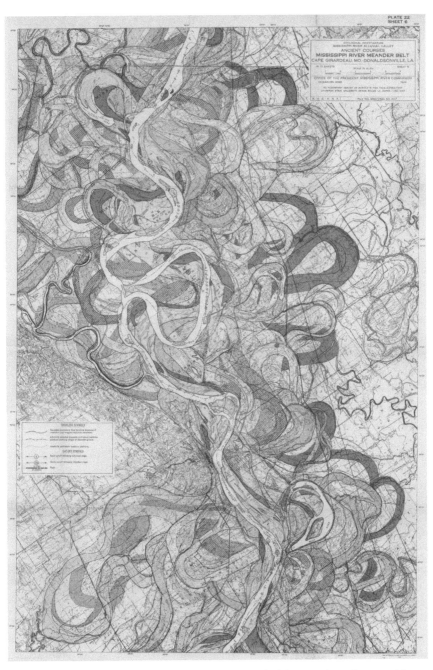

FIGURE 5. Fisk Maps. Harold Fisk, an LSU geologist, tracked the meander path maps of the entire Lower Mississippi River from Cairo, Illinois, to the Gulf of Mexico, using aerial photos, geologic samples, and hand-drawn cartographic representations dating to 1795. Image courtesy of USACE historic photo.

that influence river dynamics. The US Army Corps engineers relied, therefore, on empirical adaptation and "inductive reasoning," says Reuss.[61] River construction was reduced to a regimen of standardized steps. "When planning a project, they would ascertain the river's shape and geometry, the velocity and volume of water passing a particular point, regional geology, and the quantity and concentration of sediment."[62] But as their study moved from planning into the field, the Corps became a champion for large "scientific" projects—which were political by nature—such as canals, dams, and ports. Often the costs of large projects were only partly divulged to Congress to begin a project that would clearly require subsequent authorizations to complete. Lobbying the Corps for projects was a national pastime of elected officials from every corner of the country. Louisiana, for example, was unabashed in its flood protection requests. Such requests were made through a decentralized system of local, regional, and state harbor boards, levee commissions, and other public constituencies.

Such ambiguity about river management allowed various constituencies to advocate for solutions that served their local interests. While flood control was the main preoccupation among southern states, there were many civilian engineers, some Corps engineers, and even downstream residents who were beginning to advocate various combinations of outlets and spillways beyond a wholly "levees-only" approach. As early as the 1840s, the Louisiana State Engineer, Paul Octave Hebert, theorized that spillways would divert water only during heavy floods. Many engineers favored a controlled spillway outlet at Bonnet Carré upriver from New Orleans, which had been the site of the 1844 levee break into Lake Pontchartrain.[63]

THE SWAMPLAND ACTS

An 1849 flood swamped New Orleans when it broke through the levee at Pierre Sauvée's plantation 17 miles upriver. Within three days, floodwaters reached the French Quarter. Nervous residents on the uptown side considered severing the levee at the New Basin Canal behind the French Quarter. The idea was met with the threat of armed response by downtown residents. Three weeks later, the upper New Basin Canal collapsed anyway, which deluged 220 midcity blocks and forced the evacuation of twelve thousand residents.[64] Afterward, Congress acceded to pressure from southern constituents to pass the Swampland, or "Swamp-buster," Acts, which provided a mechanism for levee construction through land reclamation. The federal government turned over riverine bottomlands and swamps of the batture to states, which sold them to private interests. States used the money to finance levee building and flood control projects. A year later, Congress extended the program to California, Florida, Oregon, and eleven states in the Ohio-Mississippi valley. By 1909, nearly 82 million acres had passed into private hands through the swampland program, from ten cents to $1.25 per acre.[65] The swampland program also boosted the flood control cause by sponsoring Army Corps of Engineers surveys, which consistently recommended that the federal government build protective levee projects.[66]

Swampland grants financed levee lines and drained swamps in California, Arkansas, Louisiana, and Mississippi and were responsible for accelerating the growth of large-scale agriculture and forestry. This led to the formation of state levee districts, which directed construction work and had the power to sell bonds based on the future value of the previously overflowed land. Such district laws varied from state to state. They generally required a petition by the owners of the land to be drained. Once approached, a district could assess taxes in proportion to the landowners' estimated future benefits, which were determined by a board of reviews. The Swampland Acts trained capitalist landholders to organize politically. Flood control advocates were now well funded and organized to lobby the Army Corps of Engineers for flood control assistance.[67] "The creation of levee districts was, therefore, an important step for an agency that had devoted itself solely to navigation improvement work."[68] The acts allowed the federal government to encourage flood control outside of its official purview. Before 1860, less than 100,000 acres of farmland were serviced by drainage districts and municipal drainage projects. In the next forty years, the amount of acreage served by districts grew exponentially. In the last decade of the nineteenth century, more than 6 million acres of US farmland were served by drainage projects.

NEW ORLEANS AS THE LARGEST SLAVE MARKET

As the riverlands were being surveyed, de-snagged, and de-muddied, so too was the political economy of slavery and global capitalism. Demand for enslaved labor throughout the southern river basin increased as adjacent swamplands were drained and cleared for plantation agriculture.[69] In this way, Extractive Thinking was part and parcel of *racial capitalism*—a heuristic that interrogates how race and capitalism intersect in the production of racialized differences that undergird capitalist exploitation of land and labor.[70] Cheap farmland attracted thousands of white slave owners from the Atlantic seaboard to make their fortune. A settlement boom in the Delta was comparable to a second gold rush. "The second middle passage," as the historian Ira Berlin describes it, "was the central event in the lives of African American people between the American Revolution and slavery's final demise in December 1865."[71] Enslaved people were transported in forced migration overland and by sea from the older slave states to the newer cotton states and sugar plantations of South Louisiana. By 1850, a quarter of New Orleans's population had come from the North.[72] An article reprinted in the New Orleans–based *De Bow's Commercial Review* in 1858 touted the strength of local levees to protect newly reclaimed delta farmland: "We can levee successfully! We have but one outlet, the Yazoo Pass, and the levee there, the heaviest and highest in the world, has stood the flood. It stood because it was properly and securely built."[73]

The population in Mississippi and Alabama doubled between 1840 and 1860, from 179,074 whites and 195,211 enslaved to 354,000 whites and 436,631 enslaved. Cotton production more than doubled in half the time, from 194 million pounds in 1849 to 535 million pounds in 1859.[74] Some ships also continued to smuggle persons from West Africa and the Caribbean. By 1860, the number of enslaved people in Louisiana had multiplied sixteen-fold, to more than 331,000.

The expansion of plantation capitalism required clearing forests and mechanical transportation to get commodities to market in New Orleans. As American settlers began clearing the alluvial lands of the Mississippi River basin in earnest in the 1830s and 1840s, efforts to open the Mississippi River's large tributaries, such as the Red River, intensified. One of the main impediments to steamboat trade from the Mississippi into the Red River to reach into Arkansas and Texas was a massive hundred-mile logjam, the Great Red River Raft mentioned previously. A series of natural logjams in northwestern Louisiana and southwestern Arkansas, it was passable only by small boats that could snake their way through the swampy lakes. Because it impeded access and large-scale commerce, the logjam initially provided natural protection for Indigenous inhabitants, many of whom were being pushed into the Red River territory.

The Great River Raft was the result of a spring snowmelt and rainstorms. Riverbanks in the northern basin caved in, and trees fell into the channel with their roots clinging to the soil. Spring floods could carry trees that were 60 feet long downstream and jam the narrowing meander crossings. Older trees at the bottom of the clogs rotted and sank into the riverbed. "Silt collected in the interlaced branches, cementing the cottonwood and cypress trees into natural dams. Men rode horses on the largest obstructions, brushing past live trees 18 inches thick."[75] The raft creeped and cracked. "Timber at the foot, or lower end, slowly rotted away and disintegrated while yearly storms added new trees to the head." More material washed down than rotted away, which increased the raft size as it inched north. Western travelers described large inland swamps consisting of trees and driftwood that reduced "the Mighty Red River to a slow moving sludge."[76] The main channel could be clogged with an assortment of cypress, white gum, cottonwood, ash, pecan, hickory, willow, mulberry, and locust trees that were entangled with canebrakes, vines, and creeping plants. Natural dams pushed backwater into the countryside, forming chains of bayous, lakes, and wetlands that created sanctuaries for thousands of birds. Bison roamed the isolated prairies and woodlands above the raft.

The modern American eye saw not a wonderland of natural bounty but a land to be conquered and cultivated with Black and Brown bodies. Peter Custis was sent to the region by Thomas Jefferson in 1806. He described a future "paradise of America" that could be produced by cotton plantations once the raft was removed.[77] Opening it up to investment and cultivation that was spreading into South Carolina, Georgia, Tennessee, Mississippi, and Louisiana required regular steamboat service. Cotton was replacing tobacco as the main cash crop, and steam-boats would accelerate the transformation. "Steamboats were both a means and an end to transforming the environment." The forces of manifest destiny required not only the logs and "rafts" to be cleared for grand plantations, but also the Indig-enous inhabitants who were in the way. "While the raft was intact, Americans sur-rendered to nature rather than conquered it."[78] Capitalists like Capt. Henry Shreve looked upon the great raft not as a natural a priori feature of the landscape but an impediment to be removed. His snag boats began clearing the great log raft to open the Red/Atchafalaya complex to trade.

Starting in 1834, the state of Louisiana began removing river debris with the nation's only antebellum state snag service. The state Board of Public Works applied an $85,000 federal Red River appropriation to "purchase 50 good hands." It bought two skilled men the following year while liquidating a state-sponsored railroad.[79] According to the historian Aaron Hall, the Louisiana legislature pur-chased men, which it then wielded as "public hands." Their strength, intellect, health, and knowledge became instruments of statecraft to convert clogged bayous into navigable waterways that then enabled planters' enslaved laborers to turn waterlogged prairies into profitable cane and cotton fields. "Purchasing men enabled Louisiana to develop a skilled, ever-available service corps on the cheap."[80] It allowed for "flexible, year-round waterway clearing." Season after season, the enslaved became experts "sustained by a proslavery cost-benefit analysis that per-suaded Democrats and Whigs alike."[81]

Because of to its riverine swampy geography, Louisiana was particularly reliant on waterway upkeep, which generated a bureaucratic formation depen-dent on slavery-based expertise and governance on all levels of organization. The program was Louisiana's most stable and largest public-labor sector. In this way, the institution of slavery was inseparable from state building. As Hall writes, "State slavery stimulated administrative growth."[82] The state reduced expenses by placing fully enslaved crews under a white captain. Slave-operated boats included the *Crab, Experiment, General Walker, Harmanson, Agnes, Flor-ida, Pioneer, Two Friends, Franklin, Governor Hebert, Amite, Randall, Algernine,* and *Atchafalaya.* "These crafts evolved from modified steamers into custom snag boats bearing cranes, reinforcements, and double hulls to ram, yank, and cut obstructions."[83]

The public hands lived on the boats, which they kept in service through modified improvements, toolmaking, and maintenance. "With no other resi-dence, no place of rest beyond bunks or riverside camping, the enslaved men inhabited a working and living space that overlapped in ways that served state interests."[84] When the program ended in 1861, some of the men had been employed clearing bayous and rivers for years. Planters and communities peti-tioned their officials for help and in return received the labor of public hands whose training and expertise became instruments of statecraft. Slavery was not just a peculiarity. It was essential to the political economy of Louisiana to build, support, and expand its agricultural economy and to reinforce the belonging of whiteness. State legislators, governors, commissioners, engineers, and super-intendents dispatched the enslaved workforce with boats and machinery to extend a riverine network for the steamboat age. "Being a slave-master state was a collective, democratic project for Louisiana's enfranchised white pub-lic, who exercised their political power to sustain the program and receive its benefits," writes Hall.[85]

Public slavery—in solving Louisiana's riverine problem—became a *public good*. Governor André Roman declared that the answer to Louisiana's riverine gov-ernance problem lay in systematic public slavery: it was "not only the cheapest, but the only means of succeeding in a regular and permanent system of Internal Improvements" that was less expensive than white laborers and more reliable than immigrant labor. "By recasting rafted rivers and impassable bayous as problems to remedy with government slavery, the state developed on the ground and in public perception," writes Hall.[86] Naturally, then, the enslaved men experienced the vio-lent backlash of the river itself to human intervention.

- In 1834, Louisiana deployed engineers, steamboats, and fifty-seven enslaved men. Nineteen came from the interstate slave trader Isaac Franklin; 13 were dead by January 1835.
- In 1835, legislators authorized purchasing up to "two hundred able bodied negro men, none of which to exceed thirty years of age, of good character, and healthy." The service counted 88 men in 1836 before death took 7. Deadly conditions pushed the population below 70 men before officials increased "the state force" to 129 by 1850.[87]
- On June 9, 1860, the state's Internal Improvements Division sold 79 men at the City Hotel auction block. They included two engineers, four blacksmiths, three pilots, three carpenters, two cooks, and ten boat mates. Where a wilderness had existed, an engineer in 1848 recorded "extensive and highly cultivated fields, gardens and comfortable dwellings, quarters and out-houses; forming almost one continuous settlement."[88]

The public works improvements carried out by public slavery in order to open the Red River valley to plantation agriculture spurred population growth through-out the region, particularly where water navigation was opened for market deliv-ery. Louisiana's population increased from 215,739 (including 109,588 enslaved) in 1830 to 708,002 (including 331,726 enslaved) in 1860.[89]

White settlers poured into the region, dragging with them thousands of enslaved Black people, who were forced to cultivate the landscape. "Because the swift change from wilderness to cotton cultivation required the labor of thousands of slaves, a thriving interstate slave trade to the region blossomed." The city of Shreveport almost overnight served as a transfer point for agricultural products going downriver to New Orleans and for finished goods moving upriver.[90] An area that Captain Shreve described as "remote" in 1832 produced one-fifth of Louisiana's cotton crop eight years later.[91]

This settler boom placed severe pressures on Indigenous peoples who had been on the move since American expansion into western lands. There were multiple wholly distinct tribal groups living within the Red River–Ouachita watershed. The Caddo, Alabama-Coushatta, Cherokee, and Quapaw had been enjoying relative isolation. The Red River raft impediment limited their exposure and allowed relative protection to hunt and cultivate wild corn, pumpkin, and vegetables. Some Indigenous bands fled to Spanish Texas by 1801–2 in order to avoid the Americans just as they had escaped the British.[92] Others, like the Apalachee, Biloxi, Pascagoula, and Taensa, had successfully petitioned for lands between the Sabine and Trinity Rivers in Spanish Texas, but few had moved by 1802 when Louisiana was handed back to France by Spain and rumors grew that it would soon be American territory. Some began to leave in the following years, with most of them going to Texas after 1820.[93] The Caddoan speakers occupied most of northwestern Louisiana until they ceded land to the United States in 1835 and attempted to move into Texas. "After 1840, they moved en masse to the banks of the Blue River in the Kiamichi Mountains and into Texas and Mexico, leaving the area that had been their home for well over a thousand years."[94]

DELTA SURVEY

The disastrous 1849 floods that inundated most of the Lower Mississippi Valley and resulted in the 1849–50 Swampland Acts also increased congressional focus on the river itself. Floods were not the only problem. There were also sandbars. "At the mouth of the Mississippi enormous sandbars often blocked access to the Gulf of Mexico. Sometimes 50 ships waited there for the sandbars to dissipate enough to allow passage into or out of the river; the largest ships sometimes waited as long as three months." There was still no consensus on how best to control floods and improve navigation. On September 30, 1850, Congress authorized a complete survey of the lower valley from Cairo, Illinois, to the Gulf. "The aim was to discover the laws governing the Mississippi River and determine how to tame it."[95] Secretary of War Charles Magill Conrad authorized the Mississippi Delta Survey, calling on military engineers to conduct a study whose primary purpose was to discover a means to prevent flooding. After pressure from some congressmen and after conferring with President Millard Fillmore, Conrad divided the $50,000 appropriation between the army and the US Civilian Corps, each of which would issue competing surveys.

Charles Ellet would lead the civilian survey. Educated in France, he had the year before published an essay in which he described his method for allowing year-round navigation of the river: the use of reservoir basins along tributaries of the Ohio to store water during flood season that could spill back into the river during low water stages. Ellet issued his report within a year.[96] It was 150 pages and outlined four reasons that floods on the Mississippi were growing: the expansion of cultivation of farmland, which meant that forests and swamps no longer absorbed floodwaters and runoff; the extension of the levee system; the creation of cutoffs; and the lateral elongation of the river into the Gulf of Mexico. His prescription required three approaches: stronger levees, improved natural and artificial outlets, and a system of high dams and floodwater reservoirs to release excess water during low water season. His work is considered more theoretical exposition than a survey.[97]

The other report was assigned to Andrew Humphreys of the US Topographical Corps, which would merge into the Corps of Engineers in 1863. Humphreys would take eleven years to complete his report because of his assignment by Sen. Jefferson Davis to direct the Pacific Railroad Surveys, followed by multiple health-related breakdowns. Once finished, the report would make a lasting impact on national river management policy. The *Report upon the Physics and Hydraulics of the Mississippi River* was completed just months before the Civil War.[98] In the report, Humphreys and his coauthor, Lt. Henry L. Abbot, a fellow West Point alumnus, rejected reservoirs and cutoffs. It concluded that building continuous levee lines would "concentrate" the flow of the river.[99]

Humphreys's survey teams painstakingly obtained data on crossing river channels and geologic formations. They took measurements from the confluence of the Mississippi and Ohio Rivers to the Bird's Foot Delta. They studied the tributaries of the Lower Mississippi. They applied insights from geology and European hydraulics to challenge the conventional wisdom about alluvial deposits. The result, declared the *American Journal of Science*, was "one of the most profoundly scientific publications ever published by the U.S. government."[100] Their final analysis recommended maintaining all water flow within the main channel by closing natural outlets that drained water away into adjacent swamps. It recommended stronger levees. The endorsement of a "single-channel theory" tied flood control interests to navigation interests but also happened to be the most politically attractive option.[101]

Humphreys went on to serve as a Union officer during the Civil War. He was commended numerous times for his service in battle, where his stubborn tenacity earned him the moniker, "the Fighting Fool of Gettysburg," for resisting a Confederate attack. He served as chief of staff of the Army of the Potomac under Maj. Gen. George G. Meade. In November 1864, he took command of the II Corps, earning more accolades in contributing to Robert E. Lee's final surrender at Appomattox Courthouse.[102]

Humphreys was promoted to brigadier general. Sixteen months after the Civil War ended, Gen. Ulysses S. Grant appointed him to lead the Army Corps of Engineers. Ellet, meanwhile, served as a Confederate colonel during the Civil War and was mortally wounded at the Battle of Memphis. But Humphreys would soon identify a new foe in the person of James Eads, against whom he would battle for the future of the Mississippi River.

THE CORPS, POLITICS, AND SCIENCE

With the Delta Survey, the Corps of Engineers proved it was the one institution capable of gathering and analyzing data that was necessary to plan large-scale flood control programs along America's waterways. Although the Humphreys-Abbot levees-only "universal formula" later proved flawed, their report received the respect of engineers around the world. "No one could fault the authors' ambition, intelligence, and diligence. In this, Humphreys and Abbot clearly surpassed their fellow army engineers."[103]

Surveys and intervention on the rivers moved Congress toward national planning and bonded engineers to the state. "[Engineers] saw themselves as an emerging professional class whose technical expertise would command respect."[104] The Corps had become powerful because it utilized statistics and packaged information to shape government, while Congress issued shifting mandates and ambiguous legal directives. The Corps' powers of implementation extended federalism but also stirred a critique that dogs its management to this day. The rise of scientific professionalism was a gradual process. Statistical reporting, cost-benefit analysis, specialized field offices, and standardized forms and regulations were sold as the solution to partisan gridlock. The professional state was a response to the chaos that had stalled public works. But the Corps was not immune to chaos either. "The Corps, say its defenders, suffers for sins of Congress, but engineers invite the abuse by overselling their science and lavishing public money on four-color books and pamphlets that downplay the long resistance to federal projects."[105] There was seldom a time in American history, not even wartime, when the Corps worked in a political vacuum without facing stiff opposition from river organizations and bureaucracy.[106]

THE HUMPHREYS/EADS DICHOTOMY

As artfully explored by John Barry in *Rising Tide*, the disagreement over river management played out in a rivalry between two of the largest personalities in river management. Once ascendant, Andrew Humphreys met a new archrival in James Eads, who had personally experienced the alluvial floor of the river. As a river diver, he did not believe that the riverbed was made of blue clay as Humphreys asserted. He believed that a jetty system to constrain the width of the river would increase the velocity enough to scour the bed and lower the level of the river even during low stages. He also advocated for cutoffs to shorten the river's path to the Gulf and hasten the water flow.[107] Eads argued that a jetty system could also deepen the troubled mouth of the Mississippi River, whose shallow mud lumps and sandbars regularly blocked shipping lanes. He made a formal proposal to open the mouth in 1874. By then, as chief of the Army Corps of Engineers, Humphreys was developing plans for a shipping canal that would bypass the mouth. Humphreys vociferously fought, undermined, and attempted to sabotage Eads's proposal, leading Eads to put up his own money to win the project.

Humphreys believed any jetties created to increase the power of currents would be offset by tides from the Gulf. The House rejected jetties and passed Humphreys's $7.4 million canal request. The Senate, though, refused to consider it. The two bodies compromised by creating a new board of engineers.[108] Staffed with three army engineers, three civilians, and one member from the Coast Guard, the board voted 6–1 to allow Eads to put up $10 million of his own money to achieve and maintain a depth of 28 feet. Grant signed the legislation for the jetty project over the objection of Humphreys. On June 9, 1875, a steamer left from New Orleans tugging a pile driver and three flatboats of housing materials. They arrived in a cloud of biting mosquitoes at the Mississippi River's South Pass, one of three openings at the Bird's Foot, where a torrent of brown water boiled into the green expanse of the Gulf.

The new jetty system, Eads promised, would carve a channel using the river's own sediment to accomplish hydraulically what months of dredging could not. At the height of the project, eight hundred fifty workers drove several thousand lumber piles deep into the muddy banks of the pass. In less than three months, the piles "extended in a lonely curve of wood two and one-third miles into the Gulf."[109] The piles were lined with thin, flexible willow tree trunks coated with limestone-based concrete that acted as fascine mattresses. All the raw materials, the lumber, the willow trees, and sandstone, had been collected upriver and shipped to the project site. Within a year, the partially completed lining had already begun compressing the current and deepening the channel of the pass.[110] Within three years, Eads's jetty system had blown away impermanent sandbars and accomplished what levees could not. By squeezing the water, the river was able to scour its own channel. Eads proposed building modified jetties all along the lower river, making cutoffs, and temporarily confining the river with levees to help concentrate the flow and deepen the channel. In the 1870s, both Eads and Humphreys sought to force their views on others at the cost of scientific debate. Both castigated engineers who advocated for a more diversified approach to flood control, and neither tolerated dissent.

In the wake of the success of Eads's jetties, which were completed in 1879 (and remain today), his supporters proposed a bill to create a commission of civilian and military engineers independent of the Corps of Engineers, with Eads presumably as chair. Legislators in favor of such a commission saw it as an opportunity to decrease their reliance on the Corps' advice. The Mississippi River Commission (MRC) would be a seven-member body consisting of four civilian engineers and three representatives from the Army Corps of Engineers appointed by the president. The MRC would oversee future internal improvement projects on the Mississippi River and advise the Army Corps of Engineers. Humphreys opposed the bill since he felt the Corps was entirely capable of managing the Mississippi, and he retired shortly after it was established.

Although Eads became a commission member, northern critics saw some of President Hayes's early appointments to the commission as evidence that Hayes intended it to promote levees for flood control.[111] Debates again focused on the constitutionality of aid that might directly benefit private landowners rather than navigation. Private engineers attacked the Corps' competence. However, many were at least as concerned about their exclusion from public works projects. Criticism of the Corps appeared in numerous journals and periodicals.[112]

Many officials, however, continued to hold the Humphrey report in high esteem.[113] "Humphreys came to identify attacks on the report as attacks on the Corps itself." He became increasingly frustrated and defensive about emerging engineering concepts.[114] The more he was attacked, the less willing he seemed to modify his position. Tragically for the Corps, it was this inflexibility that became his main legacy rather than scientific dedication to truth.[115] It is not clear that Humphreys biased his report to give it an obvious political appeal. "Quite the contrary," according to Reuss, "he insisted on a rigorous unbiased approach to the work. However, when he finally did arrive at the levees-only policy, he firmly put his reputation behind it and defended it before critics within and outside of Congress."[116]

By the 1880s, army engineers were building flood control levees all along the Lower Mississippi River. As the levee lines became more complete, downstream residents continued to suffer. The levees-only approach was causing the river to carry a greater volume of water, and it was raising the riverbed, thus forcing engineers to construct taller levees. Any break in these larger, modern levees wreaked tremendous devastation. Yet other proposals to manage river flooding, such as opening spillway outlets into bayous, required the government to appropriate private lands, which met resistance. "Levees-only" represented a political compromise, supported by enough engineers and scientists, along with Delta landowners, to become accepted policy.

After years of organizational work, personal politicking by activists, and actions by sympathetic legislators, flood control supporters eventually won official aid for the Mississippi and Sacramento Rivers with the 1917 Ransdell-Humphreys Flood Control Act.[117] It directed the Corps of Engineers to provide levee aid for the two rivers and authorize federal payment of up to two-thirds of the cost of levee construction. Local interests remained responsible for acquiring the rights-of-way and some maintenance costs. Yet, in spring 1922, the Lower Mississippi River flooded again. The river was so high that its tributary waters flooded six Yazoo–Mississippi Delta counties. Some critics blamed the flood on the closure of the Cypress Creek Gap by the Corps of Engineers the year before. The only remaining outlet on the

Mississippi was at Old River, where Capt. Henry Shreve made his 1831 cut to the Red River. The cut would later threaten to open a permanent course for the Mississippi down the Atchafalaya River away from New Orleans and require an ongoing regime of intervention through the Old River Control Structure.

GATHERING STORM

During the winter and spring months of 1927, the Mississippi River surpassed record flood stages. Prolonged rainfall in the headwaters swelled its tributaries and increased the already elevated water levels in the Lower Mississippi. In April 1927, waters began to rise precipitously, approaching 60 feet above mean sea level. Federal levees along the Lower Mississippi began to breach. By May, floods had devastated thirty-two towns and cities and pushed the Ohio tributary backward.[118] On May 24, the river broke through Old River and sent 30-foot waters down the Atchafalaya. The breach panicked New Orleans authorities, who convinced the Corps of Engineers and the Mississippi River Commission to dynamite the levee south of the city. They used thirty-nine tons of dynamite—starting in a noontime spectacle of press, VIPs, and newsreels—to sacrifice the small rural farmers and fur trappers downstream for the good of the city. The first explosion opened a ditch 10 feet deep and 6 feet wide. Two more explosions followed to little avail. Workers used pitches and shovels. Divers set depth charges beneath the surface. Finally, the disappointed crowds went away. It would take another ten days of dynamiting to do the job. When it was over, 250,000 cubic feet per second, or 20 percent of the river's volume, poured through a hole 3,213 feet wide into St. Bernard Parish and Plaquemines Parish on its southeastern border. According to John Barry, 526 claimants filed suits against the city of New Orleans. Total claims reached $35 million. The city settled $3.9 million but then deducted $1 million for feeding and housing the claimants while they were homeless. Of the settlement, half went to the Acme Fur Company for losses of the muskrat habitat caused by the flooding. Not a single trapper received any compensation. Another $600,000 went to Louisiana Southern Railroad. The remaining $800,000 was divided among 2,809 claimants who averaged $284 for losing their homes and livelihoods. An additional 1,024 claimants received nothing, despite commitments by Louisiana governor Oramel Simpson that all victims would be compensated.[119]

Some have argued that city leaders were motivated to show investors the extent they would go to protect capital investments at the port, which was essential to the city's economy. In truth, the river had crested north of Baton Rouge the month before the dynamiting. All spring, local newspapers had played down the panic. Yet business was drying up.

Referred to as the greatest peacetime disaster in US history by Secretary of Commerce Herbert Hoover, the 1927 floods caused staggering economic losses and human suffering. Over 16 million acres in seven states were inundated, and property loss estimates varied from $236 to $363 million. Nearly 700,000 people are known to have died. Another 637,000 were left homeless. The American Red Cross, responsible for most of the relief work, provided food and shelter for more than 300,000 people in refugee camps. Black refugees were particularly harmed. In refugee camps, they were coerced to perform manual labor and prevented from fleeing the Delta because planters were afraid of losing their workforce.[120] In the aftermath of the devastation, Congress ordered the Army Corps of Engineers to examine the flood problem in a national context.

Congressional appropriations to "fix" the Mississippi River ballooned over the next half century. In 1928, Congress approved the Mississippi River and Tributaries Project, known colloquially as Project FLOOD, which gave responsibility for federal flood control projects to the Army Corps of Engineers. "The federal government now undertook the full cost of levee building and left to local levee boards only the tasks of obtaining rights-of-way and maintaining the completed levees." In 1930, Congress authorized the construction of twenty-three locks and dams for the Upper Mississippi River. In 1932, the MRC cut sixteen canals into its meander paths, shortening the distance from Memphis to Vicksburg by 170 miles, to increase the river's velocity and scour its own channel. In the New Deal spending era, river investment increased. In 1944, Congress authorized another 150 projects, totaling $750 million.[121] In all, the Corps has constructed over 300 multipurpose reservoirs, as well as thousands of miles of levees, bank revetments, and strategically placed floodgates, pumps, and upstream dams on far-flung tributaries. The federal government has enlisted billions of dollars of infrastructure to build spillways, cut through wide meander necks, seal off the remaining Louisiana marshes from its progenitor, and construct Old River Control at the confluence of the Mississippi, Red, and Atchafalaya Rivers about thirty minutes north of Baton Rouge. Entire floodplains have been walled off from the river's mouth all the way to the city of Cairo, Illinois.

The techno-politics to maintain the river in its current course requires constant maintenance, dredging, and lobbying. Project FLOOD did not quiet the cacophony of river politics in reaction to the many attempts to prevent another major flood after 1927. Such measures have both led to the seizure of private land to redirect floodways through Louisiana and launched new problems in the form of massive coastal erosion and land loss at the river's lower delta. Since the 1930s, Louisiana has lost 2,000 square miles of wetlands. Today it is not only a loss of silt and sediment that Louisiana must contend with but also the politics that produced the Father of Waters that flows through it.

Whose Paradise?

The license plate of the car ahead reads, "Sportsman's Paradise." It's newly pressed. The old plates used to say, "Bayou State." But this one sounds . . . more hopeful. The only bayous I had known were full of mosquitoes and unseen snakes and alligators. "What does it mean?" you ask your dad.

"It means that it's a good state for sportsmen," Dad says from the driver's seat.

"Sportsman?" you ask. There are not that many sports teams in central Louisiana where you were growing up. There are paved roads, traffic lights, and a mall. There is high school football and baseball. There are LSU games on television. Just beyond town, there are screened porches and camps, bonfires and barbecues, and fish fries and crawfish boils. There are lakes and rivers. There are horses. There are barns. There are cows. But you've never seen anything like a sportsman.

"What do you mean, sportsman?" you ask.

"Hunters," he says. "Duck hunting, deer hunting, fishing—you know—sportsmen."

"Hunting is a sport?"

"For some people," he says.

Never a definitive answer, which was frustrating.

4

The Birth of River Science
and Grassroots Greenwashing

To understand what Foucault called a discursive episteme, he said we must create "a history of the present" that disrupts the *truths* that have become part of the natural order of things. One quick example is Lisa Bloom's book, *Gender on Ice*, which explores how masculine gender construction and visions of empire shaped and motivated North Pole expeditions. Such imaginaries behind the construction of the pioneer explorer led to more than riveting tales of survival. They led to identity formation and policy to support more of the same imaginaries. Discourses and the truths of things are heavily dependent on the institutional formations behind knowledge production and circulation. How we come to understand the relationship between the Mississippi River and Louisiana wetlands relies on the institutional formation that produced both cultural and scientific knowledge about the river and the Louisiana coast. Our valuation of the coast for its extracted resources is part of this discursive regime. But how did this regime come to be? What were the political forces that contested and shaped our understanding? What were the economic structures surrounding this knowledge? What shaped the science—and scientific questions—that produced authoritative knowledge about them?

Science and policy on the Mississippi River had radically transformed the river from a mudscape to a consequential waterway through nineteenth-century flood control and river management. Mud was sacrificed on behalf of a rising nation; and today's river emerged from a modern understanding of what a river was intended to be. The largest river on the continent symbolized the strength and commercial potential of a nation. This vision of a commercial highway of water played out in contracts awarded to investigative teams. Levees were built, and the floods of 1927 followed. The response to the catastrophic floods was a congressionally approved infrastructural program that removed the river's mud from public view and attempted to control the height and behavior of the river that spawned an ongoing regime of infrastructural support and reinvestments that continues today.

In a pure *dialectic* fashion, this hardening infrastructure led to dire effects on the fragile coastlines of Louisiana and in turn spurred intensive investigation of river physiology and its role in producing the delta's coastal wetlands. Early studies from LSU that produced this knowledge were supported by national oil interests and the US Army Corps of Engineers—two participants with agendas that often opposed one another—while also further damaging the state's wetlands. The research programs that studied river morphology and wetland construction were indebted to and directly subsidized by the petroleum industry, the petroleum-friendly Louisiana Geologic Survey, New Deal funding, and the Army Corps of Engineers, the latter needing institutional research assistance for its ongoing project of controlling the Mississippi River through mechanical engineering principles.

The legacy of petro-dollars in Louisiana has fundamentally shaped how the state articulates its strategy for wetland management and how Louisiana residents accept the existence of oil and gas as an economic benefactor and part of the natural order of things. This confounding dilemma took shape alongside knowledge of the river's importance in delta construction. Knowledge development accompanied a period of rapid coastal erosion that coincided with the discovery of oil and gas deposits beneath large salt domes in the marshes. In fact, LSU researchers collected the very field samples that expanded their understanding of delta construction while doing contract work for the petroleum industry to survey land parcels for fossil fuel development. As they coronated the river as the marsh's progenitor, they blamed levees for reducing sediment replenishment. This, perhaps unintentionally, provided political cover for oil and gas interests whose canals, pipelines, and drilling platforms were destroying marshlands hectare by hectare.

EARLY STUDY OF LANDFORMS: BERKELEY ON THE BAYOU

We start in the early days of the LSU Department of Geology, which was officially established in 1922 by Henry V. Howe's arrival from Stanford University.[1] Dr. Howe came with a mandate from Louisiana governor John M. Parker to build a department "to train Louisiana boys for the oil industry."[2] He was charged with rebuilding a minor department that had collapsed four years earlier, "leaving only scattered heaps of rocks, minerals and fossils." Known for his enthusiasm for the subject matter, Howe attracted several students and began to lay a new foundation for a department that would be intimately tied to the state's petroleum industry.[3] Soon after joining the faculty, Howe persuaded the administration to hire his colleague, Richard Russell, who received his PhD from the University of California, Berkeley, to help develop the field of physical geography at LSU. In September 1928, Russell arrived in Baton Rouge, where he and Howe together built a major program of geology and geography, establishing the Louisiana Department of Conservation, which combined geomorphological, archaeological, and botanical reports in a single bulletin. The bulletin provided a publication venue for many of the early studies of the Mississippi River Delta.[4] Much of their research on Louisiana landforms was tied to contracts with the petroleum industry to appraise property values and land titles.[5] Howe soon made two additional hires: B. C. Craft, to train students in petroleum engineering, and Fred Kniffen from UC Berkeley, who had a strong background in cultural geography, anthropology, and geomorphology. "From this strong academic nucleus, the departments of geology, geography-anthropology, and petroleum engineering were combined in 1931 to form the School of Geology with H. V. Howe as its director."[6] Kniffen bridged anthropology and human geography, which allowed him to work with Russell to create a methodology that considered the habitation patterns of prehistoric Amerindians in response to changing river patterns. They were able to date river patterns by uncovered Indigenous artifacts. Russell's physical approach and Kniffen's archaeological analysis were a natural fit.[7] One trip that involved a visit to Larto Lake in central Louisiana resulted in the theory that the lake had once been a former channel of the Mississippi River miles from its present location.[8] Russell accompanied Kniffen on four trips into the lower river delta. "In all, Kniffen visited 44 sites that included earthen mounds, shell mounds and middens, and natural beach deposits containing pottery. He sketched mounds and bore sites and collected artifacts from the surface of sites."[9] Kniffen found multiple sites with collections of pottery: Natchez, Tunica, Caddo, Bayou Cutler, Coles Creek, Deasonville, and Marksville. Russell used Kniffen's site survey and prehistoric chronology to date the subdeltas of the Mississippi River. In a 1939 paper, "Quaternary History of LA," they concluded that Bayou Teche, where no Native American habitation artifacts were found, was the oldest Mississippi River course.[10]

Russell was able to demonstrate that a subdelta identified by the LSU archaeologist James Ford was older than the current St. Bernard subdelta. Ford, a student of Kniffen, started the archaeology program at LSU in 1937. After receiving a BA from LSU in 1936, he remained a research archaeologist there until 1946. While working on his graduate degree at the University of Michigan, Ford organized a WPA program for Louisiana, which helped create excavation sites throughout and establish an outline of the ceramic chronology of the Lower Mississippi Valley, including the Tchefuncte culture and the late prehistoric Plaquemine Culture site in West Baton Rouge Parish.[11] WPA digs also oversaw the excavation of Bayougoula in Iberville Parish, which identified the villages of Bayougoula, Mugulasha, Acolapissa, and other tribes of the late seventeenth-century period of historic contact.[12] Ford developed a timeline for the Lower Mississippi Valley, resulting in the cultural sequence: Tchefuncte—Marksville—Troyville—Coles Creek—Plaquemine—Natchez/Caddoan.[13]

The work demonstrated to Russell that three of the Native American cultures in the lower delta were relatively recent. Russell matched this evidence with marine shells close to the surface at New Orleans and the sequence of channel positions to theorize that the deltas themselves were young. His research on the geology of Plaquemines and St. Bernard Parishes also helped verify the historical meandering pattern of the river that created the Louisiana delta. His classic 1936 paper, "The Physiography of the Lower Mississippi River," combined geomorphological, archaeological, and botanical reports. He pointed out that in addition to the active delta, a sequence of abandoned deltas was in varying stages of decay in coastal Louisiana. In other words, the delta was dynamic and in various stages of formation and erosion. A delta was either in the process of growth because of the active sedimentation or in the process of decline because of a changed meander channel that was subject to natural subsidence and the stress of coastal tides. The paper argued that the delta has dominant natural levees that form the high land. The gentle slopes of these natural levees lead away from the river to marshes, swamps, and open waters. Upstream, the floodplains have tributaries; downstream, the deltas have distributaries and abandoned channels. "Meanders are present only on the floodplains where the channels encounter material deposited during the same cycle of alluviation and where the banks are lined by natural levees."[14]

The paper also introduced for the first time the concept that the weight of the sedimentary deposits of successive deltas caused local down-warping of the Earth's crust, which created a geosyncline.[15] The delta was naturally sinking under its own weight. The work was one of several contributions on delta studies published by the Louisiana Department of Conservation.

EARLY GUSHERS

By the time of Kniffen and Russell's collaboration, the nascent oil industry was forming. By 1918, Louisiana ranked sixth among oil-producing states mainly due to northern production.[16] The state issued the first coastal zone oil lease in 1921, and land development companies began acquiring huge tracts of swampland. Timber and fur companies that had exhausted their land, like Continental Land & Fur Co., were incentivized to hold onto their tracts and lease them for exploration. In fact, the oil boom promised that even land too wet for agriculture or timber had potential value for what lay below its surface in mineral rights. During the 1930s, as swamp and marshlands suddenly became valuable, legal issues of ownership arose. The state had title to navigable waterways, which hinged on the boundaries of water bodies in 1812 when Louisiana was admitted to the Union. New appraisals of property values were required. Russell's fieldwork in alluvial morphology attracted interest in soliciting his skills as an expert witness in the various land title lawsuits. Russell's extracurricular work as an expert witness on landforms sometimes earned him more money in fees than his university salary. Charles Anderson writes in his biography of Russell, "He and Howe presented evidence that won Louisiana title to extensive water bodies in southwestern Louisiana. From this activity, came the addition of the term 'Chenier,' meaning ridge of sand, to the terminology of geomorphology on Cameron and Vermilion parishes in 1935."[17]

In addition to Howe's academic responsibilities, from 1934 to 1940, he served as director of research for the Louisiana Geological Survey, which was supported by the petroleum industry. The survey brought oil money into the department that supported more faculty and graduate students. Oil money subsidized fieldwork and made possible the geologic mapping of the state's parishes. And it established bulletin publications of the State Survey. Howe personally authored or coauthored the first eight parish bulletins. The papers emphasized the importance of the thick, elongate, sedimentary sequence paralleling the coast, which is the main source of Louisiana's petroleum. Howe also developed concepts of salt dome growth and recognized the significance of subsidence under deltaic loading, Pleistocene terrace formation, and the Quaternary deltaic history of coastal Louisiana.[18] The petroleum lobby wanted to expand the geologic survey under LSU's management with proposed legislation to triple their fees with new drilling permits. However, Howe and Russell resisted expansion pressures. They cited the difficulty of training personnel to interpret geologic evidence in the densely vegetated and muddy coast. "In order to establish precise locations for necessary boring and land surveyors, they had to cut trails through the swamps or walk miles on unstable floating marsh. In some cases, botanists, chemists, and other specialists were included in the field parties."[19]

In 1937, Russell was given the first Wallace A. Atwood Award by the Association of American Geographers. His groundbreaking report established him as one of the leading geomorphologists in America.[20] Russell served on the Committee on Geophysics and Geography for the Department of Defense and served as adviser to the Office of Naval Research. During World War II, German U-boats prowled Gulf waters to disrupt Allied shipping lanes from the Mississippi River. After the war, in 1949, Russell was urged by army and navy officers to help improve the "trafficability" of vessels throughout the coastal complex. This offer came as he was named dean of the Graduate School at LSU. With the assistance of James P. Morgan, he presented a proposal to the Geography Branch of the Office of Naval Research to study the ability of large vessels to navigate the shallow, muddy Louisiana coastal marshes. This study led to the establishment in 1954 of the Coastal Studies Institute with Russell as director,[21] as well as the dredging of navigational canals. Today there are ten major navigational canals connecting the Gulf of Mexico to inland Louisiana ports. Studies indicate that the presence of these canals allows salt water to intrude into the freshwater marshes, especially during storm surges. Dredging of straight canals through channels that previously meandered accelerates the speed of storm surge and tidal action, causing destruction of the healthy wetlands. In addition, canals with high spoil bank edges—where the dredged mud is stacked along the bank of the canal—disrupt the hydrology of wetlands. This results in deterioration of marshes and ultimately loss of land to open water.[22]

FISK'S ARRIVAL

Like the Louisiana Geologic Survey, the Army Corps of Engineers was interested in commandeering research labor from the LSU department. They were seeking a better understanding of the physics and morphology of the Mississippi River. The Corps in 1941 hired Harold Fisk, who had arrived at LSU six years earlier.[23] With expertise in volcanic rocks, Fisk began making discoveries in central Louisiana and formulated an explanation of Quaternary deposits. For the Corps, Fisk undertook a comprehensive geologic study of the entire alluvial valley of the Lower Mississippi. Nothing of such magnitude had previously been attempted.

The Fisk study, completed in 1944, included a summary of the valley's characteristics, chronology, and historical evolution. The investigation provided a glimpse into not only the major factors that led to the establishment of the river's modern course but also what may shape the river's future behavior. Fisk's team used detailed topographic maps, aerial photography, and historical accounts of the river valley, which included narratives from sixteenth-century Spanish explorers, to help identify abandoned courses of the Mississippi River and its tributaries. They also incorporated data from sixteen thousand soil borings.[24] Fisk revised the original sequence proposed by Russell. They worked out some of the details of the development of the deltaic plain. In later work, LSU researchers further revised this sequence.[25] Fisk left LSU in 1948 to join the oil industry. He became chief of the Geologic Research Section of the Humble Oil and Refining Company in Houston and stayed on as a consultant to the Army Corps of Engineers.[26] The Fisk study had a profound impact on the geologic understanding of the Mississippi River Valley and would drive Mississippi River engineering for decades.

Along with the report were several volumes of multicolored, detailed topographic and geologic maps that set a new standard for geologic illustrations. These maps trace significant river course changes over the past two thousand years. For instance, the river has taken at least three different routes through Louisiana to the Gulf of Mexico; its present course through New Orleans dates only to around 650 years ago, although more recent studies suggest its present course may be younger.[27] During his service to the Army Corps of Engineers, Fisk was apparently shocked at the militant culture of the Corps, which operated under strict standard procedures. "For example, whenever they began a new construction project, such as a levee setback, they made borings at very regular intervals without considering the surface geology of the area."[28] Fisk and his team early on established definite relationships between types of sediments such as gravel, sand, silt, and clay and the common, different depositional environments of the floodplain such as natural levees, low flood basins, abandoned channels, and point bars—all of which was later recognized as a major breakthrough in classic sedimentology. As Fisk busily mapped abandoned river courses and stages of the current meander belt, the Army Corps of Engineers was studying alluvial river models in the laboratory. Residents and researchers, meanwhile, began to note the disappearance of Louisiana's marshlands.[29]

NEW UNDERSTANDING OF COASTAL PROCESSES

During this period of erosion and in the following decades, a range of new research revealed the river's mechanism of delta construction and ultimately the impacts that both the Army Corps of Engineers' ever-growing system of levees to hold the river in a single channel and the growing labyrinth of oil and gas canals were having on the lower delta marshlands. The early research on mounds, villages, and the meandering Mississippi and its deltas began to open new approaches to teams of geomorphologists, geologists, and archaeologists. "A wedding of cultural and physical science took place, resulting in what Kniffen in 1958—with a premonition of what was to happen nationwide in the 1970s— called the birth of cultural ecology." Later studies followed this direction, with particular attention to the cyclic nature of delta growth.[30] Prior to the 1980s, it was universally thought that the coastal marshlands were in symbiotic balance. But geologists began to gather and examine historical aerial photographs and other evidence that revealed a very different story. Coastal erosion in Louisiana was accelerating.

In 1964, the Louisiana geologist Sherwood Gagliano began a series of studies on dynamic formation and erosion of river deltas, as well as on the ecological devastation caused by disrupting the Mississippi's natural sedimentation cycle. His 1964 paper, "Cyclic Sedimentation in the Mississippi River Delta," looked at the natural effects of river deltas when rivers changed course and began building new deltas. The abandoned delta was essentially "starved" of nourishment and began a coastal retreat that led to its eventual inundation by seawater. Gagliano examined subdeltas around southeastern Louisiana that were no longer main channels of the Mississippi River. His work was joined by a raft of studies by natural scientists that coincided with the emergence of the environmental movement in the late 1960s. Evidence mounted that correlated land loss to both the management of the Mississippi River and the promiscuous oil and gas exploration and drilling.

In 1971, the Louisiana Legislature established the Louisiana Advisory Commission on Coastal and Marine Resources. The same year, a pair of studies was released by R. H. Chabreck that quantified land disappearance in the coastal zone at a rate of 16.5 miles a year. Titled "Ponds and Lakes of the Louisiana Coastal Marshes and Their Value to Fish and Wildlife," the study relied on a helicopter survey of marsh vegetation and soils. Chabreck sampled quarter-mile intervals over a study area of more than 12,000 miles, followed by a second survey of 20,488 miles.

These studies were compiled in the survey, *Hydrological and Geologic Studies of Coastal Louisiana*, in which he began pushing for a more coordinated state-led response to coastal land loss.

Mapping studies by coastal scientists continued through the 1970s, showing that land loss had nearly doubled by the end of the decade. Evidence pointed overwhelmingly to man-made factors, and a coastal restoration movement began to take hold: "The apparent causes of the high rates of land loss include the harnessing of the Mississippi River by levees and control structures which reduce tendencies toward natural diversion and funnel valuable sediments to deep, offshore waters. Additional factors include canal dredging and accelerated subsidence related to mineral extraction, both of which are often associated with saltwater intrusion. The net effect is a rapidly accelerating man-induced transgression of a major coastal system."[31]

Essentially, practices to control the river, which were thought to be either benign or beneficial for two centuries, suddenly became tied to an alarming rate of coastal erosion. Around 1970, Gagliano began another series of landmark observations that measured the impact of levees on protected land. He found that levees were contributing to a separate problem related to land loss. Their very weight was causing subsidence. He identified three separate processes: (1) by cutting off the natural flow of water, levees were essentially starving the root systems of the land inside the protection rings; (2) the levees themselves were causing the land to sink, since they resided on mud—consisting of organic peats and soft clays, silts, and sediments—that was too soft to bear their weight; and (3) the organic peat on developed land inside the levees shrank as the water was drained and pumped outside of the levee system. "Furthermore," Gagliano wrote, "when dried, they shrink appreciably with a volume reduction of as much as 50 percent or more in the peats." The clays fared a little better, with a maximum reduction of 10 to 15 percent if dried completely. But when exposed to oxidizing conditions, through diking and draining, they sank further.[32]

Drained back-swamp areas—like many of the postwar neighborhoods developed in New Orleans and adjacent suburbs—were simply poor landforms to build on. They could sink by several feet. Other practices such as using fires for clearing stumps or accidental fires could smolder in the peat for months and cause further subsidence. "The net effect is that the surface of the newly reclaimed land may be lowered five feet or more below sea level within a decade after drainage. If the drained area has ponds and small lakes, the effects may be even more drastic."[33] Starting in 1973, Gagliano began advocating for controlled diversions along the Mississippi River and the need to deploy the river's sediment load to replenish the marshlands through a dynamic management plan.[34] But dedicated funding for restoration projects and supporting applied research would not become a reality for another decade.

In the late 1970s, the US Fish and Wildlife Service and the US Bureau of Land Management contracted with Gagliano and his firm, Coastal Environments Inc.

(CEI), to produce a series of habitat maps for use in planning for Outer Continental Shelf (OCS) oil and gas development and the potential impact of future expansion on fisheries. It resulted in the "Mississippi Delta Plain Region Habitat Mapping Study" in 1980, under the direction of Karen Wicker, also of CEI. Wicker developed a methodology for establishing hydrological units and analyzing habitat maps to determine change. Wicker's study used data imagery from aerial photos taken in 1978, which she overlaid with previous maps to produce a series of new maps of the entire Louisiana coast, which consists of two plains: the Mississippi River Delta Plain between the Bird's Foot Delta south of New Orleans to Vermilion Bay in Iberia Parish; and the adjacent Chenier Plain in southwestern Louisiana. The Wicker habitat maps produced dramatic results: 465,000 acres of marsh had been lost in the Mississippi Delta between 1955 and 1978, with an average loss of 32.3 square miles per year (20,600 acres). But the Chenier Plain in southwestern Louisiana was sinking faster. The total loss came to 25,000 acres, or 39 square miles, a year, twice the rate identified by coastal scientists ten years before.[35]

ROOT CANALS AND EROSION

As the methodologies for studying land loss improved, scientists using aerial photographs started recognizing another correlation to the land loss phenomenon: the state's weblike labyrinth of canals—thousands of miles of them. By comparing aerial maps, researchers in the 1970s began plotting the conversion of wetlands to open water as a result of canal construction. In 1983, a study by Eugene Turner using Wicker's 1980 mapping study and earlier data argued that canals were causing local submersion of wetlands. The density of navigation and oil pipeline canals in a given geographic area was directly proportional to high rates of land loss, especially in younger, abandoned deltas, such as Terrebonne and Plaquemines Parishes, southeast of New Orleans. The question became not whether canals directly contributed to land loss but by how much. An emerging theory about *indirect* land erosion associated with canals would become a major point of contention and controversy that plunged restoration science into a political quagmire for decades in Louisiana.

In basic terms, land loss meant converting wetlands to open water. Turner's study went on to investigate how canals might be harmful to marshes, which bled into questions of horticulture, hydraulics, and invasive species. Canals opened saltwater channels into fresh and brackish marshes. This had some effect on plant and marine ecology. Canals entrapped water behind dredged spoil ridges stacked along the edges of canals during the excavation process. Turner argued that impounded water altered salinity and decreased nutrient, organic matter, and sediment exchange. It changed vegetation composition and reduced vegetation productivity. Land loss, his research found, was directly proportional to canal density in certain areas: "Canals also allowed salt water from the Gulf tides to seep into the coastal interior, well beyond the protective natural barriers, killing cypress swamps and freshwater marsh vegetation and increasing subsidence in these areas."[36] Gagliano had written in 1973 that canals threatened to "seriously upset natural circulation patterns and water chemistry." Canal excavation had made "the petroleum industry the greatest wetland canal builder." Gagliano cited the practice of adding new channels without any effort to refill the old ones. "When a canal is cut, it often becomes a permanent feature." Single canals became webs of canals that coalesce into small lakes and bays.[37] Eventually, pipelines connected offshore discoveries to onshore processing and transportation facilities. "Flow lines and pipelines, connecting the fields infrastructure, sat in canal bottoms or on their banks."[38]

But canals also signaled oil field activity, which included surface-level disposal of brines that contained several times higher levels of salinity than seawater. And then, what of the oil pipelines themselves? Louisiana soils are particularly caustic and degrade industry infrastructure. Subterranean leaks and ruptures caused unknown havoc on the fragile ecology. A changing plant ecology was then less resilient to opportunistic species of beetles, nutrias, birds, and other creatures. Still, an even thornier question related to the fact that this growing web of canals resided on private property, often owned by absentee owners who received oil and gas royalties. They had a vested interest in resisting any sorts of claims of culpability in coastal erosion by oil and gas interests.[39]

CANAL EXCAVATION BY INDUSTRY:
SUGAR, TIMBER, AND OIL

Canals provided access to the resources in the marsh-swamp complex. Historical maps indicate that marsh and swamp drainage ditches were excavated as early as 1720. Such watercourses helped drain agricultural land and extract cypress timber. The earliest available maps indicate drainage channels were the first artificial waterways used in resource exploitation. They were built by the French for both land drainage and access channels. Coastal dwellers cut small, narrow marsh passageways called pirogue trails or traînasse trapping ditches for quick access to their traps and an easy way to move furs and other supplies. Hence, the most intensive trapping networks prior to the twentieth century developed in muskrat and nutria feeding areas. By 1915, as muskrat fur became fashionable, pelts increased in value and trappers extended their hunting ditches into muskrat habitat. Trapping canals connected fields to camps, where the animals were skinned and dried. During this early period, a good trapper could catch up to two hundred muskrats a day.[40] A string of men working a section of ditch cost about $10 to $15 per mile. The channels were small but effective in the trapper's efforts to catch muskrat, mink, otter, raccoon, and nutria, which were introduced from Argentina in 1937 and are now an invasive species.

But the most intensive were the canals for logging, petroleum, and transportation. In order to remove cypress from the swamp, the lumber industry built navigable connecting canals through which pull-boats dragged large stands of timber that left still-visible scars. The sugar and timber industries collectively decimated the Louisiana cypress swamps as land reclamation projects using levee and drainage programs coincided with the rise in the mid-nineteenth century of major sugar plantations. Wetland reclamation techniques were based on a system of levees, internal drains, and pumping plants. Once the area to be reclaimed was defined, a large dredge-boat canal at least 25 feet wide and 5 feet deep was constructed around the project's perimeter. In building these canals, the dredged spoil was used as a protective levee with a height of 5 to 6 feet and a top width up to 12 feet. When the boundary canals and levees were completed, the internal drainage network was constructed.[41] The political ecology led to profit-taking twice over: denuding and selling the cypress forests to clear land for planting monocrop sugarcane.

The sugar industry and its subsequent demand for wood-powered steam in the mid-nineteenth century further decimated the state's old growth timberlands. The introduction of the railroads provided direct access to logging—while disrupting the ecology of the marsh through levee rail beds that impounded water. "By the 1920s, operators had removed 4.3 million acres of timber from the state, an area about the size of New Jersey."[42] As the cypress and other virgin stands were depleted and the practice ended, only the canals remained as evidence. However, the collapse of the cypress industry coincided with discoveries of petroleum and natural gas along the coast, which would amplify the extractive commodification of the coastal forests and marshes. After 1930, oil companies began to use canals in their work, which required cutting larger passageways through already traversed marshes and thereby changing the transportation patterns along the coast. These patterns were further altered with the completion of the Gulf Intracoastal Waterway through the marshes in the 1940s. "Mudboats" were utilized to excavate ditches west of Vermilion Bay but were not employed east of the Bay. The Larose ditch digger, introduced around 1933, could move through the deltaic plain by cutting trails in the floating vegetation and highly organic soils.[43] A floating vessel 20 feet long and 5.5 feet wide, it was powered by an inboard engine. It housed two bow-mounted 36-inch rotating cutting blades designed like an airplane propeller capable of excavating a 6-foot-wide, 6-inch-deep channel. The propeller cut the vegetation and at the same time pulled the boat through the marsh. The machine could cut a traînasse in five inches of water.

HABITAT AND ECONOMIC LOSSES

In practice, habitat loss associated with industrial canal and pipeline dredging— as well as routine bleeding of oil and brine into the surrounding estuary—was a well-known occurrence for much of the twentieth century. As early as 1913, Percy Viosca, a Tulane-educated conservationist who was head of the Louisiana Department of Conservation, later renamed the Louisiana Department of Wildlife

and Fisheries (LDWF), began sounding the alarm through two separate administrations of state government. By 1925, Viosca was proselytizing about disappearing
wetlands and saltwater intrusions from navigation and drainage canals as well as
brine water dumping from wells. "Man-made modifications in Louisiana wetlands,
which are changing the conditions of existence from its very foundations, are the
result of flood protection, deforestation, deepening channels, and the cutting of
navigation and drainage canals," he argued.[44] And an agency biologist reported
in 1940, "Through the digging of canals[,] good muskrat country can be readily
and quickly ruined."[45] It wasn't just muskrats being affected. Oyster beds were also
being fouled, which was a problem because oysters represented a growing industry. The state began leasing oyster seed grounds along the coast in the 1910s, and by
1960, it had leased over 70,000 acres. The number climbed to 400,000 acres by the
end of the century. Gulf oysters accounted for two-thirds of the nation's domestic
oyster supply.[46] At a 1953 conference on oil and gas impacts, James McConnell, the
LDWF's oyster and water bottoms chief, provided empirical observations about
the disruptions caused by spoil ridges. The industrial-like arrangement of tank
batteries, processing facilities, and disposal areas was incompatible with the marsh
setting. Houck quotes McConnel: "Everyone should recognize that there are
other very old industries here . . . that are now being seriously affected by these
mineral operations."[47] The LDWF consistently found thick layers of floating oil
in the coastal estuaries and oil-related drilling mud that settled at water bottoms
where oysters bred. "They are going to destroy an industry to build another industry," C. H. Brookshine, an LDFW commissioner, stated in 1955.[48] A year later,
the 1956–57 biennial report by the Louisiana Wildlife and Fisheries Commission noted "drastic increases in salinity" and "rapid deterioration" of the marshes
around Barataria Bay Waterway. At that point, at least two wells a day were being
drilled along the coast.[49]

    In a 1959 report on drilling in the Rockefeller Wildlife Refuge and Game
Preserve, an area supposedly demonstrating the compatibility of oil and wildlife for many years, an LDWF researcher noted that the more than 20 miles of
access canals there, within a few years, had enlarged by 20 percent. In the early
1960s, a report out of Texas concluded that canal dredging could also be a reason
for increased salinity in the Louisiana marshes. Van Lopik of LSU echoed the
findings of LDWF scientists that "many oil company canals, with their flanking
spoil banks, cross the marsh giving rise to changes of drainage, and hence, vegetation. Thus, relatively minor modifications in marshland drainage may create
many unforeseen problems." By 1971, Lyle St. Amant of the LDWF was even more
emphatic, pronouncing the canal effects as for the most part "irreversible and permanent" and representing a "true ecological upheaval." He described wastewater
discharge and pollution as "rampant and uncontrolled," reasoning that "the coastal
region of the state was a virtual trackless wilderness" that allowed "oil waste, leakage, sludge, and other materials [to be] dumped into the marshes and bays without

FIGURE 6. Derrick in the Marsh. This undated photo shows an oil derrick surrounded by a waste pit in Terrebonne Parish. A million oil and gas wells have been permitted in Louisiana since the 1901 gusher in Jennings. Oil and gas activities caused more than 70 percent of land loss in the Barataria-Terrebonne basin. Image courtesy of the Historic New Orleans Collection, © Douglas Baz and Charles H. Traub, 2019.0362.93.

regulation or control."[50] By 1973, even the Army Corps of Engineers recognized that "onshore pipeline construction may cause irretrievable marshland loss."[51] More studies detailed the mechanics and extent of the damage. In 1983, an article by Turner cited the work of more than twenty professionals in the field, each investigating one aspect of canal damage or another. Turner's team called for limits on dredging permits by the state and the Corps of Engineers, as well as new construction techniques and requirements for backfilling new and existing canals. "It wasn't arm waving. You had data across the coast," said Turner. "We had maps of change. We had maps of little ponds, big ponds, straight lines. Finally, people were looking at this as a whole system. So, it fit. It was a whole coastal view. There were differences in there, but, my God, every place on the coast was losing land."[52]

Once upon a time, canal spoil ridges were seen as beneficial for hunting and fishing. Hunters would burn the grass on one end of the muddy strip and shoot the animals retreating toward them. Some landowners would build and lease camps on spoil banks. In 1971, a group applying for a 6-mile pipeline through the marsh in southern Louisiana at Pecan Island was approved by the interstate regulatory Federal Power Commission, which found that whatever disruption caused by the canals would be more than offset by improved deer habitat and access to trapping grounds provided by the spoil banks.[53] Some landowners believed that the spoil banks would protect them from hurricanes. "In the background, there was always the possibility that oil companies would want to re-access the wells for more production."[54] Turner and Donald Cahoon conducted another major study in 1987, "Causes of Wetland Loss in the Coastal Central Gulf of Mexico," which made the first serious attempt to quantify the land loss indirectly attributable to pipeline construction in the wetlands. It found that the oil pipelines directly contributed to 4 to 5 percent of land loss from 1955 to 1978, but a much higher percentage of indirect effects was caused by "water logging." When water is unable to naturally drain back toward the Gulf, the salt water begins to change the chemical and biological conditions of the marsh soil. Over a short time, the marsh vegetation will deteriorate and soils will oxidize, leading to another cause of land loss—subsidence. Over time, internal ponds will enlarge. Pipelines that ran into the marsh from OCS drilling, they estimated, were responsible for 8 to 17 percent of land loss.

Turner and Cahoon analyzed the few backfilled canals in the region and determined that backfilling reduced direct impacts of the canals by as much as a third.[55] From 1979 to 1980, the Louisiana Offshore Oil Port excavated a 5-mile pipeline canal, which was then backfilled. Field studies done in 1985, a relatively short time later, showed a third of the spoil areas to be between 23 percent and 75 percent covered with renewed marsh. Although the backfill had not yet fully restored original conditions, the corrective resulted in shallow water areas with higher habitat value for fish and wildlife compared to unfilled canals.[56]

The implication that federally permitted pipelines in Louisiana caused significant land loss led the US Minerals Management Service (MMS) to initiate coastal impact studies of its own.[57] The MMS published "Pipeline Impacts on Wetlands: Final Environmental Assessment," which found that during the period 1951–82, the government approved 72,870 miles of pipeline rights-of-way on the Outer Continental Shelf in the Gulf of Mexico. Approximately 130 of these pipelines made landfall on the Louisiana coast. The report looked at five pipelines built between 1978 and 1984 and determined that pipelines and canals did have major impacts on marsh vegetation. Fast forward to 2009, when the same agency, renamed the US Bureau of Ocean Energy Management (BOEM), said that construction of OCS-related pipelines can cause intense habitat changes and conversion to open water locally. It found that the practice of direct dredging opened areas to saltwater intrusion and that spoil banks altered flooding patterns.

But impacts associated with specific pipeline canals varied. Some pipelines contributed to habitat loss and others didn't, depending on the quality of mitigation applied and the kind of habitat that it crossed. A report by Johnston, Cahoon, and La Peyre stated, "Our analysis also suggests that the cumulative effect of hundreds of pipelines contributes to regional trends in land loss."[58] In 2010, a backfilling test on old oil and gas access canals in the Jean Lafitte National Park's Barataria Unit compared success rates in two sections, one restored simply by pushing in the spoil banks and the other by adding soil from other sources to hasten the process. Both demonstrated progress. Somewhat unexpectedly, the test area where the spoil banks were simply pushed in recovered at the same rate as the section with the soil enhancement.

BACKFILLING: THE QUICK DEATH OF A CONCEPT

Turner argued that backfilling old canals costs a fraction of the proposed river diversions. Several times, he said, the Environmental Protection Agency (EPA) argued for pilot projects to backfill canals. His study with Walter Sikora in 1985 looked at all known examples of backfilling from 1979 to 1984 and concluded that, where properly done, the technique restored natural hydrology and began the process of infill of the open canals. In 1987, Turner and C. Neill did a follow-up study of some thirty sites, confirming further progress.[59] In 1994 and 2004, the ten- and twenty-year marks, more follow-up studies showed more progress still. In 2005, an analysis by Turner's graduate student, Joel Baustian, showed wetland recovery in 65 percent of the spoil areas and 25 percent of the formerly open canals.[60]

The state Department of Natural Resources (DNR) Coastal Management Division made one attempt to implement backfilling. Prompted by the Sikora and Turner study in 1985, program administrator, Joel Lindsey, forwarded it up the chain to the secretary of DNR, which regulates the oil and gas industry, with a memorandum summarizing its conclusions that "even partial backfilling" of a canal was beneficial, creating a shallow lake occupied by marsh-typical organisms and reducing "water logging" in the adjacent wetlands, which allowed revegetation of marsh plants to occur.[61] The fieldwork was characterized as "done in a professional manner," "outstanding," and "excellent." Continental Land & Fur, a major royalty owner, opposed it, as did the state DWF, which concluded that "several recommendations stated in the report cannot be justified" and requested that "the report not be published in a final form until our concerns are addressed."[62]

According to Len Bahr, a former state coastal scientist-turned-critic, the state DNR had an incestuous relationship with the industry that corrupted its mission of overseeing oil and gas exploration and production. In 1989, DNR was also given responsibility for implementing a small program of coastal restoration managed by the Coastal Wetlands Authority. This was the predecessor of the Coastal Protection and Wetlands Authority (CPRA), which today oversees billions of dollars of restoration and protection spending—subsidized by oil and gas royalties.[63] "The fact that DNR was given responsibility for overseeing the conflicting missions of coastal restoration and energy production is no accident of course, having been masterminded by the energy industry," said Bahr, who served as the coastal adviser to Gov. Mike Foster. "Purveyors of oil and gas were clearly fearful that coastal management and restoration might step on their lucrative financial toes."[64] According to Bahr, it was ironic that the need to restore the coast is the result of damage caused by the fossil fuel industry. But as we will see in the next chapter, the oil and gas industry managed to reposition itself not as a culprit of land loss but as its victim.

Meanwhile, the industry contested the applicability of a Louisiana law that mandated restoration of wetlands to their preexisting condition, called Section 705. Backfilling, the industry argued, was only required "upon cessation of use for navigation purposes." A company might at some time want to go back and work over a rig or a drill in a different direction. And in the meantime, the canals were "navigated" regularly by Louisiana fishermen. The arguments stymied backfilling within the DNR's Coastal Management Division. Consequently, DNR announced a temporary moratorium on Section 705, which had rarely been implemented.[65] The Turner and Sikora study was sent for review to LSU's Center for Wetlands Resources, which found it "inconclusive." The moratorium was never lifted. Nor are there any projects or studies in the state's current master plan for coastal restoration to test backfilling. Today, the thirty projects surveyed by Turner and his colleagues over the past four decades represent almost the entire sum of all backfilling done in the Louisiana coastal zone. That's fewer than 10 restored miles of more than 14,000 miles of canals from the Texas border to Mississippi. Following up in 2014, a master's thesis in environmental management by a Duke University student examined backfilling potential coastwide and its projected costs. It found over 100,000 acres damaged by canal banks, of which nearly half had the necessary features for success. "Based on the highest cost per acre estimate available," this acreage could be successfully backfilled for $8.7 million. By comparison, the state's 2017 iteration of the master plan outlined a suite of techniques (none of them backfilling) to restore an area slightly less than 20,000 acres at an estimated $3 billion. This was nearly four times the cost per acre using techniques less proven than simply pushing in the spoil banks and letting nature do the rest. Backfilling canals with material from existing spoil banks would restore some of the natural hydrological function at a low cost-benefit ratio. These small-scale restoration projects through backfilling would have allowed the marsh to be "stitched" back together relatively cheaply. "There are quite a few thousand abandoned canals. If they were officially abandoned, [the state] could have them backfilled, but they didn't do it. There are a lot more that are practically abandoned but not legally abandoned."[66]

According to Turner, the very act of backfilling assigns the culpability of oil and gas production to marsh erosion: "If you're going to do it, it means someone has to be blamed." The industry has resisted taking responsibility, and therefore backfilling was never embedded in the larger social framework of restoration.[67]

Today another major hindrance to backfilling is the fact that many of the spoil ridges have submerged over time so that there is no ready dirt to push back into the canal, according to critics of the approach.[68] Turner and his team argue that spoil banks alone often do not completely fill the canal with sediment. "However, this does not detract from backfilling as a viable restoration technique, because the canal becomes shallower and provides excellent habitat for a variety of wildlife."[69] As recently as 2018, Turner and Giovanna McClenachan argued that the abandoned canals connecting to the estimated 27,483 plugged and officially abandoned oil wells in the fourteen coastal parishes as officially labeled by DNR provided ample opportunities for a program of spoil bank infill and monitoring. "The total length of spoil banks in 2017 was long enough to cross the Louisiana coast east-to-west 79 times with a spoil bank height up to 3–10 times the natural tidal range."[70] Just dragging down the remaining material from the bank back to the canal could be an inexpensive long-term strategy. "The absence of a State or Federal backfilling program is a huge, missed opportunity to 1) conduct cost-effective restoration at a relatively low cost, and, 2) conduct systematic restoration monitoring and hypothesis testing that advances knowledge and improves the efficacy of future attempts."[71] The price of backfilling all canals would be $335 million dollars, or 0.67 percent of the state's master plan for restoration. It's a pittance of the profit from extracting the oil and gas below over the past century.[72]

SUBSIDENCE FROM DRILLING ITSELF

Between 1900 and 2017, the state permitted 76,247 oil and gas wells in the fourteen coastal parishes. Wetland destruction in these oil and gas fields occurred quickly. Some erosion and subsidence are natural, such as geologic faulting, sediment compaction, long-term delta lobe cycle, variability in river discharge, tidal exchange, wave erosion, and weather. While the more than 30,000 kilometers of canals dredged in the marshes are known for causing dramatic wetland loss, due to cumulative effects of altered surface hydrology, there was another correlation between deep well extraction and the disappearance of the coast. In the early 2000s, researchers began studying the impacts of drilling and extraction itself.

Decades of fluid withdrawal from oil and gas reservoirs, some believed, had increased subsidence rates in localized areas. Robert Morton, a geologist with the United States Geological Survey (USGS), analyzed what he called subsidence "hotspots" in Terrebonne Parish. He pointed to a correlation between drilled wells and wetland loss in marshy areas. According to Morton, an increasing amount of subsidence in these hotspots was directly attributable to the removal of oil and gas during the same period. Morton had worked as a petroleum geologist for a major oil company with field assignments both offshore and in Lafourche Parish, where his later studies would center. He had witnessed subsidence firsthand, noting at times how pipe casing collapsed. He did not expect his findings for the USGS to be dramatic or controversial. He and two colleagues reported on hydrocarbon production and resulting pressure losses in several large South Louisiana fields. Production showed large spikes, peaking in the 1970s, while pressure in the reservoirs fell, ultimately to near-zero, which is when the surface began to sink. The highest subsidence rates closely tracked the maximum rates of fluid extraction. Each of these fields had pumped out as much as 920 billion cubic feet of gas, 55 million barrels of oil, and 87 million barrels of brines and related waters, which were very big numbers. The report concluded, "The primary factor causing accelerated interior wetland loss in south central Louisiana between the 1950s and 1970s was accelerated subsidence and probably fault reactivation induced by rapid, large volume production of hydrocarbons (primarily gas) and formation water."[73]

By 2015, approximately 100 trillion cubic feet of natural gas and 12 billion barrels of oil had been extracted from the Louisiana coastal zone. The brines and produced waters that came up with them at least equal the figure for oil. "That removing this colossal volume of material will impact the surface above is supported by the best evidence available from home and abroad, throw in a pinch of common sense," writes the environmental law professor Oliver Houck.[74] Morton had to rely on analogical studies in coastal Texas and other coastal locations because of a lack of available data in Louisiana. As he and Bernier wrote, "Despite numerous field studies around the world since the 1920s and acknowledgment by the petroleum industry that hydrocarbon production can induce subsidence," the presence of this same phenomenon in the Mississippi Delta region "has been largely ignored."[75]

Oil and gas drilling has been associated with significant subsidence elsewhere. A well-known example is Long Beach, California, where two production fields were linked to substantial drops in the land above. Long Beach was once known as the "Sinking City" after 3.7 billion barrels were extracted from the Wilmington Oil Field, creating a 20-square-mile "subsidence bowl" of up to 29 feet deep around the Port of Long Beach and the coastal strand of the City of Long Beach. In the 1950s, water injection was shown to repressure the oil formations, stop the underground compaction as well as surface subsidence, and increase oil recovery—which ultimately led to the California Subsidence Act in 1958, which requires that well operators use water injection to repressurize wells.[76] By 1962, operators spent over $100 million on projects that included a massive repressurization program using injected salt water to reduce and in some cases reverse subsidence.[77] Subsidence episodes occurred in Venezuela's famed Lake Maracaibo and sites in Russia, Indonesia, Malaysia, and the Norwegian North Sea, "causing concern for platform safety." Experiences in the Netherlands have led to regulations requiring oil

FIGURE 7. Pipelines. Signs warning against anchoring or dredging dot the coastal marshlands. Thousands of pipelines, both functioning and abandoned, litter the coastal zone and damage fishing boats. Canal dredging and spills have decimated the once-robust marshlands that buffered communities from seasonal storms and hurricanes. Photo courtesy of Kerry Maloney.

companies there to routinely monitor and report rates of subsidence relating to extraction as they go forward.[78]

In Louisiana, petroleum lies in layers of sand pressed under layers of mud and caps of salt. The sand grains themselves are irregular, packed together like jacks in a box and buffered by the petroleum and brines. Pumping out these fluids reduces the total mass below ground. It also reduces the pressure of the formation that kept the roof up. And it removes the buffer fluids that keep the sand grains apart, which now jam more tightly together. The result is that the strata above begin to sink. It may also depressurize formations along fault lines, triggering shifts. The shallower the wells, "the more localized and dramatic these effects." The impacts of deeper wells are less pronounced, but their impacts may extend widely.[79]

Morton also theorized that deep well withdrawal may trigger fault activity among the two fault lines that crisscross the lower portion of the state. That research was controversial. Critics argued that oil and gas wells at 17,000 feet below the surface in Louisiana are much deeper than the analogous sites he considered in Galveston and California. Most of the subsidence in Louisiana, they argued, is closer to the surface and likely has more to do with organic compaction rather than depressurized wells or drilling-caused fault activity. Yet few follow-up studies have been pursued. "Morton was one of the few in the wilderness asking such questions," says Houck.[80] Making the science nearly impossible to advance provides cover for industry defenders and skeptics who claim that there is not enough data yet to support the claims.

Pointing to a fifty-year-old mandate in California that oil and gas companies reinject wells to repressurize them, Houck says, "In every other state, that is common practice."[81] Many researchers were not prepared to accept Morton's conclusions, arguing that correlation does not mean causation. "He spent his career looking for the smoking gun and never found it," according to Denise Reed, former chief scientist at the Water Institute of the Gulf.[82] Other geologists were saying there wasn't enough data. "In brief, we were data short and interested-in-getting-it short," writes Houck. "A sweep of related literature published ten years later stated that "'in the absence of more direct studies,'" the impacts of subsurface oil and gas extraction "'may never be proven.'" The evidence, it concluded, while suggestive, "'remains circumstantial.'"[83]

In a 2004 article, Reed and Lee Wilson coauthored an article that acknowledged Morton's work on subsidence and down-faulting but added the studies were in their infancy. "In some areas fault movements associated with these withdrawals appear to have resulted in a tripling of subsidence rates," they said.[84] In some hotspot areas of land loss, marsh sediments had sunk by more than a meter below their natural elevation. This pointed to subsidence at rates much higher than the few millimeters per year associated solely with the compaction of deltaic sediments.[85] In 2011, Alex Kolker of the Louisiana Universities Marine Consortium (LUMCON) presented a new method for calculating subsidence rates and found that these rates do indeed fluctuate in relation to fluid withdrawal. Onshore oil production in the state was 114 million barrels in 1945, soared to 437 million barrels in 1968, and then declined to 55.5 million in 2005—which, in sum, tracked onshore subsidence rates directly. "Taken together," Kolker concluded, "these findings point to a tight coupling between fluid withdrawal, subsidence rates, and wetland loss."[86] Extraction was by no means the sole cause, but it increasingly seemed to be a significant one. In 2014, investigations by Chandong Chang and associates at Stanford University discovered that subsidence continued after production had ended. Fluids were apparently leaking back into production cavities from adjoining areas. The first blitz of withdrawal lowered surfaces by up to 3.5 inches, followed by another potential 3.5 inches in succeeding years.[87]

But more inquiries have gained steam. A 2020 paper by John Day and others pointed directly to depressurized wells and subsidence: a deflated core pressure "induces subsidence and fault activation—especially when the production rate is high." They added that subsidence can continue "for decades" even after most of the oil and gas has been produced, "resulting in subsidence over much of an oil field that can be greater than surface subsidence due to altered hydrology." They also pointed to canals and ponding effects as well as accidental spills and intentional releases of oil and extracted brine water toxic to the area's ecology.[88]

An estimated two million barrels of brine water per day were discharged in coastal wetlands from seven hundred sites. The brine pulled up from oil wells contains toxic materials—such as benzene, ethyl benzene, xylene, and radium—that are then channeled through the canals into the surrounding marshes.[89] "This water is a mixture of either liquid or gaseous hydrocarbons, high salinity (up to 300 ppt) water, dissolved and suspended solids such as sand or silt, and injected fluids and additives associated with exploration and production activities and it is toxic to many estuarine organisms including vegetation and fauna."[90] Spilled oil is also toxic to estuarine organisms.

Meanwhile, Patricia Persaud, an LSU geophysicist, is currently working on research in fault-activated earthquakes in Louisiana that are caused by oil and gas fracking in northern Louisiana, as well as depleted salt domes in South Louisiana. Her work had just begun at the time of this writing.[91] But any findings will likely be disputed by industry. For example, in April 2019, a petrogeologist with the Louisiana Geologic Survey and president of the New Orleans Geological Society penned an op-ed lambasting the City of New Orleans's decision to join a lawsuit with southeastern Louisiana parishes against six oil and gas companies for the loss of the city's buffering wetlands. The geologist-cum-lobbyist argued that the suit takes a punitive approach to an industry that has helpfully shared 3D seismic data with universities, whose early research has found "many if not most" of wetland losses are directly associated with geologic faults. The study, he said, found that most of these faults extended to the surface, and several of them "correspond to abrupt shifts from emergent wetlands to fully submerged areas of open water."[92] In other words, he said, spontaneous movement of the faults is causing the wetlands to submerge, not the onslaught of extraction of oil and gas, disposal of toxic brines, prodigious canal excavation, and associated habitat mortality by the industry.[93]

THE BIRTH OF A GRASSROOTS RESPONSE

As residents of southeastern Louisiana witnessed the alarming disappearance of their surrounding landscape, a small but growing group of wetland advocates and conservationists in the 1980s began promoting awareness of the coastal crisis. But, like the researchers themselves, they were fragmented and divided on key issues. Intense scientific and political disagreements over the causes of wetland subsidence and erosion played out through various strategic plans and efforts to address the many stressors destroying the ecology of the area.

While some researchers focused on the actions by the Army Corps of Engineers to manage the Mississippi River, others pointed to the impacts of the oil and gas industry's canals and hydrocarbon withdrawal. Yet it seems just as plausible that they all worked in tandem. The academic community held its first major wetland conference in 1981 to identify the cause of loss and recommend options. The Coastal Erosion and Wetland Modification in Louisiana Conference affirmed that human activity had "disturbed natural processes" that had for thousands of years maintained an ecological balance between accretion and subsidence.[94] But scientists, community faith leaders, and environmentalists, painstakingly documenting the disappearance of coastal wetlands, ran headlong into misinformation campaigns and complicit legislators that muddied the water over causes of coastal land loss. Any findings that oil and gas exploration was tied to erosion were actively resisted by the energy lobby and state lawmakers. They pointed instead to the Army Corps of Engineers river levees. This despite the fact that research on navigation and pipeline canals used the same principles of water hydrology as the levee's theory, which asserted that elevated ridges like spoil banks and levees disrupt the sheeting flow of sediment deposition.

That same year, the state legislature held a special session and passed Act 41, which created the Coastal Environmental Protection Trust Fund with $35 million to fight erosion. The move coincided with a groundbreaking study in 1981 in which Gagliano quantified the loss of coastal wetlands that was directly tied to the management and leveeing of the Mississippi River.[95] Using comparisons of black-and-white aerial photographs and color infrared imagery taken at five periods from 1890 to 1978, he contextualized the rate of land loss and habitat change within the Mississippi Deltaic Plain, arguing that seven thousand years of land production by the river had been reversed starting in the late nineteenth century and accelerating in the twentieth. Such land loss rates progressed from 6.7 square miles a year in 1937 to a projected 29.4 square miles a year in 1980. The greatest loss occurred in the wetlands and barrier islands. Natural-levee ridges were also disappearing at a very high rate. Gagliano worked with the state legislature's natural resource committee to develop a restoration project list, and in the early 1980s, the legislature called for the creation of a coastal master plan. It charged the petroleum-friendly Louisiana Geological Survey to develop a ten-year plan and oversee restoration activities.[96] It funded research and a set of pilot restoration projects to demonstrate the effectiveness of various restoration techniques. The state also used the trust fund to provide its share of the cost for the Army Corps of Engineers–led freshwater diversion projects. But the project was fraught with mismanagement and lack of oversight. By 1987, only a portion of the $35 million trust fund had been spent, with most of it going to independent studies. No master plan materialized.

The great oil bust of 1980 left the South Louisiana extractive economy decimated. In May 1982, newly elected governor, Dave Treen, introduced the controversial Coastal Wetlands Environmental Levy to tax transportation of oil and gas production that moved through pipelines across state wetlands. Supporters cited research by Turner and his colleagues on the harm of canals and argued that the tax would provide reasonable compensation for the environmental cost of building pipelines. But Treen, who was the first Republican governor since Reconstruction, became a cautionary tale for environmental advocates. He targeted the industry, which responded in kind. In mass mailings, CEOs urged shareholders, employees, vendors, and landowners receiving oil and gas royalties to inform elected officials about the potential economic impacts of the bill on the state's leading industry. The mailer said it threatened increased energy prices, loss of oil field jobs, loss of state revenue, and reduced incentives for exploration. Treen's bill failed, as did his bid for a second term.[97]

Maps produced by the US Geologic Survey and the US Army Corps of Engineers throughout the decade convinced the public of the need for restoration. In the mid-1980s, a new citizen-directed initiative began to coalesce in the southern parish of Terrebonne, where wetlands accounted for 70 percent of the parish's landmass. This disappearance began to alarm the Catholic church and the impact of land loss on its parishioners. In the early 1980s, parish leaders launched an educational campaign with brochures, billboards, and classes to inform citizens. After Hurricane Juan ripped through the parish in 1985, community leaders saw firsthand how the loss of storm-buffering barrier islands and marshes led to increased storm surges and flooding onshore. Moreover, the introduction of marsh management projects on private property using weirs to regulate water flow, which was a restoration effort supported by the state, led to closures of fishing grounds.

OPPOSITION FROM OIL AND GAS

As the decade of the 1980s wound down, a consensus began to emerge, particularly among coastal experts, that the environmental cost of the oil industry had been substantially greater than previously estimated. Jim Tripp, general counsel of the Environmental Defense Fund (EDF), with experience litigating damages from navigation canals, teamed up with Oliver Houck, who was then a member of the National Wildlife Federation and the southern Louisiana archdiocese, along with a number of coastal scientists and researchers who had been studying the area for decades, to write a citizens' plan for restoration.[98] They called themselves the Coalition to Restore Coastal Louisiana (CRCL), which brought many of the groups and opposing viewpoints under a unified agenda.

The CRCL is today the state's oldest active environmental organization. Its capstone report, finalized in 1989, titled, "Here Today and Gone Tomorrow," recommended nineteen action steps for reversing coastal erosion. They included building freshwater diversions from the Mississippi River, bringing regulatory pressure to backfill petroleum canals, establishing a pipeline user fee, establishing a restoration management office in state government, and phasing out new canal construction in the marshes.[99] Some of the recommendations are now part of the state's "Comprehensive Master Plan for Coastal Restoration": using sediment and freshwater from the Mississippi River to slow land loss; pumping dredged materials from the river channel into coastal marshes; and stabilizing barrier islands through vegetation, natural processes, and beach nourishment. Other recommendations targeted the industry, such as developing "alternative means of access to

oil and gas sites within the coastal zone" and "marsh restoration by means of plugging and backfilling strategic canals." There would also be a prioritized schedule
for backfilling abandoned or little used canals.[100] The citizens' report concluded
that oil and gas production and construction of navigation and access canals were
major causes of subsidence: "While any single oil and gas canal . . . may have only
a minor effect, the cumulative impact of these canals on the coastal zone is devastating."[101] The plan was endorsed by a diverse group, including Catholic Social
Services, the League of Women Voters, the Natural Resources Defense Council, a number of local chambers of commerce, the National Wildlife Federation,
the Orleans Audubon Society, the Greater New Orleans Tourist and Convention
Commission, and the United Houma Nation.[102]

In the report, the authors asserted that 40 percent of the nation's wetlands
were under threat and were receiving little national support. "We need to think
more boldly, agree more collectively, and act more swiftly if we hope to retain
more than a few museums of marsh along the Gulf of Mexico," they stated.[103]
A letter from Bishop Boudreaux to the Houma-Thibodaux Diocese stated, "We are
morally obligated, as stewards of God's gifts, to protect and restore our coastal wetlands."[104] At the time, new GIS imagery data were providing convincing evidence
connecting Louisiana wetland losses to a labyrinth of access and pipeline canals.
Studies by the Corps of Engineers, the MMS, and the EPA all confirmed significant industry impacts. Even the *Wall Street Journal* published a series of articles
on Louisiana and the oil and gas industry, the third of which was captioned, "Oil's
Legacy: Louisiana Marshlands, Laced with Oil Canals, Are Rapidly Vanishing."[105]
The CRCL's recommendations struck a nerve with the oil and gas interests. Instead
of accepting these findings, the oil and gas lobby fought the citizens' coastal plan.
"As they had done since the beginning of the crisis, oil and gas companies and
their political supporters joined big landowners in resisting efforts to impose regulatory oversight."[106] They challenged the science, even arguing that some oil and
gas impacts were exaggerated, temporary, or even beneficial.[107]

Nationally, oil and gas corporations were simultaneously challenging new federal wetland protections across the board. In 1989, sensing new federal regulations in the wings, the oil giants Exxon, Shell, Conoco, Texaco, BP America, and
Arco Alaska teamed up with mining and real estate companies to form a lobby
ironically called the National Wetlands Coalition, which successfully lobbied two
Louisiana congressmen, Jimmy Hayes and Billy Tauzin, to introduce the Comprehensive Wetlands Conservation and Management Act of 1995, which removed the
EPA entirely from wetlands protection.[108] A blitzkrieg followed. The Louisiana-
based energy lobbying firm, Louisiana Mid-Continental Oil and Gas Association,
funded a report from three LSU geologists; it minimized the impacts of canals,
claiming they caused less than 10 percent of erosion and did not account for
any off-site impacts. Continental Land & Fur Company, with its large lease hold-
ings in fast-disappearing Terrebonne Parish, warned that more environmental

regulation would send the oil and gas sector off to "look for new places to explore."[109] Louisiana Land & Exploration Company, the largest independent oil producer in the United States, went one step further. As a rebuke to the CRCL citizens' plan, "Here Today and Gone Tomorrow," it launched a public relations campaign featuring a film, *Countdown on the Coast*, which roundly blamed the Army Corps of Engineers' Mississippi River levees for coastal erosion. Several experts were interviewed. No mention was made of the pipelines and canals.

However, the Coalition to Restore Coastal Louisiana managed to effect a series of state and federal laws that would eventually result in the passage of Act 6 by the Louisiana Legislature in 1989. The year before, Louisiana received federal recognition of its coastal wetland crisis when Congress authorized the Barataria-Terrebonne National Estuary Program (BTNEP) "for the purpose of protecting and restoring the 4.2 million acres of wetlands in the Barataria-Terrebonne estuary, one of the most ecologically productive and fastest disappearing landmasses on earth."[110] The BTNEP initiative created a conference of local stakeholders—local oystermen, fishermen, shrimpers, scientists, educators, citizens, environmental groups, and oil and gas interests—which produced fifty-one separate action steps to restore the local ecosystem.

It also provided momentum for the next plan: the 1990 Coastal Wetland Planning, Preservation, and Restoration Act (CWPPRA), which is colloquially known as the Breaux Act after its sponsor, then US senator John Breaux (D-LA). The act authorized $40 million a year for restoration projects and planning and called for the development of a comprehensive plan within three years—which resulted in the Louisiana Coastal Wetlands Restoration Plan in 1993. It called for a basin-by-basin planning approach within the nine hydraulic basins across the coastal area. The authors attributed 30 percent of the land losses to natural causes and 70 percent to man-made activities such as oil and gas extraction, saying that these activities may have triggered fault movements, as well as river levees, canals, and spoil banks and invasive species such as nutrias—all of which change the hydrology of the marsh.[111] While the funds were relatively modest, the Breaux Act codified that the state's wetlands were disappearing, and it was largely industry's fault. It also is credited with establishing a multiagency task force to begin restoration actions, which would continue to be the model going forward.

Basin teams nominated projects that were selected by a task force. The CWPPRA also provided a monitoring program for the first twenty years of each project. The initiative resulted in two large-scale freshwater reintroductions at Caernarvon and Davis Pond. The act was originally funded from small engine fuel taxes from the Highway Trust Fund. Its funding was reauthorized four separate times, but it was having little impact on Louisiana's coast. Modeling forecasted that the CWPPRA would prevent less than 20 percent of land loss by 2050 and that the state should expect to lose more than 600,000 acres of wetlands in fifty years.[112] While the CWPPRA had been intended to provide a comprehensive approach to restoration, it lacked region-wide strategies for better integration and for technical and policy review.[113] The projects under the CWPPRA, which succeeded in "preserving, creating or restoring 75,000 acres by end of decade,"[114] represented a proverbial finger in the dike. Louisiana needed a comprehensive plan.[115]

In the late 1990s, Chip Groat, who ran the Center for Coastal Energy and Environmental Resources at LSU, which was friendly with the petroleum industry, urged more big-picture solutions. What would be needed was not only a government-legislative response but also advocacy groups for wetland protection and restoration and a financial commitment on a massive scale. In 1996, Governor Foster committed to add state general funds to the coastal restoration trust fund to match all available federal funds. Mark Davis, director of the CRCL, wrote to Senator Breaux in 1997 that the CWPPRA lacked a clear vision of what kind of restored coastline it would produce and lacked a clear strategy for getting there. The state Department of Natural Resources and the Corps of Engineers disagreed on which agency would have control over contracts and project designs for the CWPPRA. Other issues included property rights disputes, interagency squabbling, and permit and construction delays. Meanwhile, the oystermen filed a precedent-setting lawsuit against the State of Louisiana over economic damages created by the Caernarvon freshwater diversion project because of a desalinization effect on their seeding grounds. This foregrounded political dissension in the master planning diversion projects a decade later.

An outside panel led by Donald Boesch from the University of Maryland found that the CWPPRA did not have enough broad-based support. They published a report titled "Scientific Assessment of Coastal Wetland Loss, Restoration and Management in Louisiana" that argued for balancing private land rights with greater public interests.[116] And in 1999, a study of a barrier island restoration was projecting land loss into the future even with all the proposed Breaux Act projects being implemented. In response, a CWPPRA task force of federal agencies and the State of Louisiana sponsored an eighteen-month study conducted by members of academia, private industry, and local, state, and federal agencies to develop a strategic plan to save the Louisiana coast, which culminated in "Coast 2050: Toward a Sustainable Coastal Louisiana" in 1998.

The "Coast 2050" plan outlined an ecosystem view of restoration and environmental management for what would be needed to maintain "essential ecological processes" over the next fifty years. Rather than a project-specific approach, it considered what the system needed to be sustainable. It recognized Gagliano's "environmental blueprint" for the coast, which called for a defensive and offensive approach, suggesting that some areas were not restorable. "Coast 2050" laid out the consensus of geologic research that most of the land in coastal Louisiana was built by deltas of the Mississippi River or by Mississippi River sediments entering the coastal mud stream. Barrier islands and sandy shorelines developed as waves, and coastal currents eroded and reworked sediments to build beaches and barrier islands. Maintaining the landscape required these and other processes. Soil-building processes would be vital to maintaining the system.

"Coast 2050" was also the first coastal restoration plan to anticipate the role of sea level rise on the coastal delta. Natural processes of sediment compaction and gradual sea level rise, it argued, submerge marsh plants and swamp forests, unless the soil can build up to compensate and maintain a high enough elevation for plants and trees to survive.[117] The plan took direct aim not only at river levees but also at canals that provided water access to drilling sites and their associated spoil banks: "Navigation channels and canals dredged for oil and gas extraction have dramatically altered the hydrology of the coastal area. North-south channels and canals brought saltwater into fresh marshes where the salinity and sulfides killed the vegetation. Canals also increased tidal processes that impacted the marsh by increasing erosion. East- west canals impeded sheet flow, ponded water on the marsh, and led to stress and eventual loss. Jetties at the mouth of the Mississippi River directed sediment into deep waters of the gulf."[118]

The plan specifically called for cutting gaps into spoil banks to release entrapped water, and it included Gagliano's Third Delta Conveyance Channel from Donald-sonville to Barataria to create a third delta. All twenty parishes in the coastal region adopted resolutions supporting the plan. The report priced what it considered a sufficient restoration program at a tenfold increase in the funding provided by the Breaux Act. "Coast 2050" outlined seventy-seven restoration strategies needed to protect 449,250 acres of coastal wetlands. In addition, it established that the natural geomorphic and ecological processes that had created the coast were impaired and that reestablishment of these processes was essential. "Coast 2050" contained two important differences from previous coastal planning efforts in Louisiana. It focused on meeting strategic goals rather than listing projects, and it took a regional view of the interventions that were needed.

The Coalition to Restore Coastal Louisiana, meanwhile, was lobbying for federal support to help pay for the plan's estimated $14 billion price tag. The organization created a national network, called "Restore America's Estuaries," to advocate for national wetlands recognition with the goal of restoring one million acres of estuarine habitat, half of which would be in Louisiana. And they explicitly urged the passage of "Coast 2050." In 2000, the coalition published a companion report, "No Time to Lose," that framed the loss of Louisiana's wetlands as an economic loss to the nation.

Once adopted by the Louisiana Legislature, "Coast 2050" required a massive infusion of funding. The state pinned its efforts to a long-standing grievance with the federal government on offshore federal royalty collections from oil. Louisiana lawmakers for years had argued that they were shouldering most of the burden of pipeline infrastructure without fair compensation. The state has also had a long-standing objection to the boundary line that the federal government has recognized since the 1930s.[119] Louisiana claims that its boundary line begins out at its barrier islands, but the court disagreed: "The Case established the boundaries of Texas and Florida at three marine leagues (10.3 geographic miles) off of their respective coastlines, while limiting the boundaries of Louisiana, Mississippi, and Alabama to only three geographical miles."[120] The court refused to draw the state's 1812 boundary starting at the barrier islands, essentially ignoring the state's unique geography. Through the 1970s and 1980s, the state lost a string of cases.

There are nearly two hundred Outer Continental Shelf pipelines that come ashore through canals and a half-dozen navigation channels built through the coastal marshes.[121] US senator Mary Landrieu (D-LA), who was friendly with the energy industry, targeted federal royalty collections. Landrieu introduced the Conservation and Reinvestment Act (CARA) of 1999. She argued that Louisiana had supported 90 percent of offshore development in the Gulf for more than fifty years and benefited from decades of economic activity but had "not received appropriate compensation for the use of its land and the environmental impacts of this production."[122] While the state received a 50-50 royalty share on oil and gas extraction within its legal boundary, it received a tiny portion of royalties between 3 and 6 miles offshore and virtually nothing beyond the 6-mile range, where most oil and gas activity in the Gulf of Mexico has taken place for decades.[123] "These areas and their fragile environments in Louisiana were sacrificed long ago for the benefit of industry investment and development," Landrieu said. "I intend to ensure that these areas will be ignored no longer."[124] The act would have boosted Louisiana's annual share of offshore revenues to about $200 million for fifteen years. Jack Caldwell of the Department of Natural Resources cited the Houma Navigation Channel as an example of a federal waterway built mainly to service the OCS that caused erosion of several square miles of land in south Terrebonne Parish over three decades. "In addition, the Louisiana coast is crisscrossed by 14,000 miles of pipelines," he said.[125] In the past, Louisiana's energy lobby had steadfastly denied the long-term impacts from dredging and drilling. But CARA did not increase royalties paid by the industry. It instead asked for a larger share of existing collections. With CARA, Louisiana was not claiming new boundary recognition but rather compensation for local costs of providing a national good.

Congressional support for the bill, however, began to wane in early 2000. Anti-drilling proponents believed that revenue sharing might stimulate additional offshore drilling. In a scathing letter, Robert Szabo, a Louisiana lobbyist, wrote to the US Senate Committee on Natural Resources, "Let me state clearly that the foundation for this bill has been, from the very beginning, Louisiana's need to restore our coast due to its unique value both to the nation and our state." He further declared that energy production from the federal OCS had generated "substantial costs" of environmental damages and infrastructure expenditures "that are either not being addressed or are being funded by the State of Louisiana and our parish governments."[126] In spring 2000, Congress took up the legislation along with an environmental bill for Florida's Everglades called the Comprehensive Everglades Restoration Plan (CERP), which was a twenty-year, $7.8 billion federal request.[127] Landrieu's amendment failed and with it a federal partnership. But Florida's bill passed. A pair of Louisiana coastal planners attributed Florida's success to a "linchpin issue" that bound disparate groups behind a common message and shared commitment. Florida's linchpin issue rested on the municipal water supply for South Florida's 20 million people. Louisiana's linchpin issue was different. "While the loss of so much physical habitat would be dire, environmental concerns alone are not sufficient to warrant the billions needed for comprehensive restoration," the planners said.[128] So how to convince Congress that Louisiana's coast—similar to Florida's coast in terms of size, rate of disappearance, and ecological inventory— was important?

Between the 1780s and 1980s, Florida and Louisiana each lost about 46 percent of their wetland acreage, or 9.3 and 7.4 million acres, respectively.[129] The authors lauded Florida's success in assembling the necessary stakeholders to publicize its erosion problem to win federal money. They questioned how Louisiana could create an identity for itself commensurate with the Everglades. Louisiana's wetlands, they said, lacked an identity. Florida's success was traced to its social infrastructure, political will, and history of activism that had begun in the 1920s. In 1994, the state had established the Governor's Commission for Sustainable South Florida, a panel of prominent industry and environmental leaders that built the political infrastructure for Everglades restoration.[130] Two years later, Congress passed the 1996 Water Resources Development Act (WRDA), which authorized the Army Corps of Engineers and the State of Florida to reevaluate a midcentury Corps project to provide water and flood control for cities and farms in southern and central Florida. This reevaluation led to the development of the Comprehensive Everglades Restoration Program (CERP) under the subsequent water resources act in 2000. Its primary goal was to return the hydrology of the Everglades to a more "natural" pattern. The $7.8 billion authorized under the plan was added to $3.2 billion already dedicated to Everglades restoration efforts since 1983. The new CERP contained more than sixty project features and was projected to create 217,000 acres of new reservoirs and wetlands.[131] When Everglades environmental activism was starting up in the early twentieth century, Louisiana was opening its wetlands to oil and gas development. "In 1923, seismic exploration technology was introduced to the region, and a decade later the LCZ (Louisiana Coastal Zone) was bustling with exploration and production."[132] After the Great Mississippi Flood of 1927, Congress directed the Corps of Engineers to construct a fortified and fully contiguous levee system along the Lower Mississippi River, which effectively cut off all sediment distribution from Louisiana's marshes. "Louisiana's appeal for restoration funding will be predicated on a host of concerns, but the linchpin issues are likely to be fisheries and petroleum infrastructure," the authors wrote. While half of the Everglades are in a national park, most of the Louisiana wetlands support commercial enterprises. "The relevant question is: *Can Louisiana convince the national interest that a 'working coast' is worth saving?*"[133]

Jim Trip of the EDF, who in the 1980s had helped form the CRCL, which ultimately led to "Coast 2050," suggested coastal advocates approach his old friend and prep school classmate, the New Orleans banker King Milling, who had deep roots in Louisiana landownership and oil and gas interests.[134] Milling was the president of Whitney Bank. Tripp and Mark Davis, the new head of CRCL, with a background in real estate development, appealed to Milling on his own terms. Whitney Bank's "collateral" of oil and gas infrastructure was disappearing into the sea.[135] Milling soon became the public face for the coalition and Louisiana coastal restoration writ large. Houck writes, "He spoke well, looked the part, and was patently sincere. He saw no conflict between saving his coast and protecting his industry. They were one and the same thing."[136]

GREENWASHING AMERICA'S WETLAND

In 2001, Republican governor Mike Foster formed the Committee on the Future of Coastal Louisiana, which in February 2002 submitted its report, "Saving Coastal Louisiana: A National Treasure, Recommendations for Implementing an Expanded Coastal Restoration Program." Milling chaired the new Governor's Advisory Commission. Also in 2001, Governor Foster organized a major coastal summit in Baton Rouge. At it, Milling declared the cost of coastal erosion should be told in dollars, commercial impact, and cultural values. "Oil and gas platforms and facilities, including pipelines[,] . . . will have to be either rebuilt or totally replaced," he said. On August 27, 2002, Governor Foster announced a campaign to increase national awareness of the state's dramatic coastal land loss: America's WETLAND: Campaign to Save Coastal Louisiana. It was funded by a $3 million donation from Shell Corp. "Although the entire nation depends on Louisiana's coastal wetlands for its energy production, seafood harvest, leading port system and wildlife habitat, very few people know they even exist," said Foster.[137] Milling became the spokesman for America's WETLAND Foundation. "He stated his conviction early and often: coastal stakeholders needed to form a new band of brothers and fight towards a common objective: securing federal (public) funding to restore the zone," Houck writes.[138]

What followed was an organized, industry-led public relations campaign. America's WETLAND partnered with Marmillion & Co., a national strategic communications firm led by a Louisiana native. A media buy was committed by TIME for KIDS to "develop educational and youth-focused materials." An educational video premiered at the 2002 Southern Governor's Association Conference in New Orleans stressing the importance of America's wetlands to the nation's energy and economic security.[139] Two days after the governor's presentation, Tripp's Environmental Defense Fund praised the America's WETLAND campaign as an important step toward "informing Americans about the value of vast but threatened coastal wetlands created by the Mississippi River." The effort to restore the coast would focus squarely on river sediment and the past practices of the US Army Corps of Engineers to levee the river, not on curtailing commercial or oil activity in the marsh itself. "Instead of being dumped off the continental shelf, river sediment should be diverted and used to rebuild wetlands," Tripp said in EDF's release. "We support the Governor's efforts to raise awareness about the plight of the wetlands and the federal funding needed to develop and implement a comprehensive, science-based restoration plan."[140]

Tripp and EDF recruited the National Wildlife Federation and the National Audubon Society to the campaign.[141] America's WETLAND sponsored international wetland science summits, organized congressional briefings, and recruited corporate sponsors. A successful campaign, they said, "will require that Louisianans speak with a unified voice and exhibit a strong commitment to paying the state's share of restoration costs."[142] At an early commission meeting in 2003, Tripp announced that "the environmental community and the energy industry must be partners as one part of creating the political will" for coastal restoration. The president of Shell Chemical echoed, "We must realize that we have been part of the problem and that we can be part of the solution." Essentially, oil and gas would fund the America's WETLAND Foundation campaign.[143]

The WETLAND group focused its energies through a campaign called "America's Energy Coast," which issued a publication called "A Region at Risk" on the nation's vulnerable energy infrastructure. The main highway to reach Port Fourchon—a major hub at the edge of the Louisiana coast that services offshore energy platforms—was vulnerable to environmental threats. "If broken by storms, floods or further erosion, it can disrupt the flow of goods and services that are the key to fueling America."[144] Senator Landrieu said, "When we lose resources so vital to our national security, it's as if we're under attack. We should respond accordingly. We would not allow a foreign power to threaten our land without a fight. Therefore, we should not allow a less obvious, but equally threatening power to take our land away."[145]

The campaign was intended to appear as a grassroots movement that would convince Congress to increase Louisiana's share of federal royalties from offshore wells. But perhaps even more important, the campaign also aimed to expand the OCS exploration to pay for coastal restoration. That would require lifting a twenty-five-year moratorium on drilling that had protected 90 percent of the OCS. As late as April 2005, Landrieu was publicly vowing to expand drilling and get a better royalty deal with the federal government. At the time, current law generally gave producing states 27 percent of revenues from production 3 to 6 miles offshore, which had to be shared equally with all states hosting pipelines, while revenues from drilling farther out went entirely to the federal Treasury. Jason Theriot explains, "The talking points for Louisiana politicians and coastal advocates had clearly shifted from solely protecting environmental resources to preserving the coast for America's energy and economic needs."[146]

As gas prices started to rise in 2005, Landrieu tried again to revive CARA. But her effort stalled after the George W. Bush administration balked at giving up federal royalties and environmentalists joined with Florida officials to oppose opening the OCS. But later that summer in 2005 something else happened: Hurricane Katrina made landfall.

Interlude

The Arrivals

They started in the eighteenth century: my father's family emigrated from England to Virginia prior to the American Revolution. As one of the less-renowned strains, our family line migrated from Richmond, Virginia, to South Carolina, before eventually landing on the Red River outside of Colfax, Louisiana, in 1879. My mother's side of the family emigrated from France to New Orleans in 1830. Her great-great-grandfather, Napoleon Joseph Frémaux, was sent back to the Collège Louis-le-Grand in France for his education and then returned to Louisiana as a civil engineer. He was also a gifted cartographer and artist. Frémaux became a naturalized citizen in 1855 and shortened his name to Léon.

I was raised on the typical boosterism of family lore without much of the racial and extractive realities that undergirded it. Both of my great-great-grandfathers served as Confederate officers in the Civil War, which was fought to maintain slavery. My grandfather on my father's side was a cotton planter, owned a company store, and undoubtedly employed Black sharecroppers. He was a quiet man whose intimidating silence, I was told, belied incessant worry about the weather and the harvest. He was eventually forced to default on the terms of a loan and lost the family land in the early 1980s.

My maternal grandfather arrived in Louisiana in 1932 to work for the booming Standard Oil Co. in Baton Rouge. A self-described Pennsylvania Dutchman, he received his master's in chemical engineering from the Massachusetts Institute of Technology during the Great Depression and caught a steamer to Baton Rouge for a job at the refinery. He met my grandmother when she was an eighteen-year-old coed at LSU. He eventually became president of Esso International, later renamed Exxon, before losing most of his eyesight to macular degeneration and retiring early to my grandmother's birthplace in the southwestern Louisiana town of Crowley, whose square-mile city grid was platted in 1887 by Leon V. Frémaux, the son of the aforementioned civil engineer.

As I wake to the humid morning and hear the trilling of insects and birdsong at a writer's retreat to finish this book about the economic, racial, and environmental history of this place, I have to consider my own debt to the colliding histories that brought so many people to this delta. Walking the river in Plaquemines Parish, I see four Vietnamese fishermen hunched on an Igloo ice chest by a pair of buckets; some hard-hat workers waiting at the Port Ship West Bank landing for a transport to one of the oil platforms offshore; the seaman on the bulk carriers waiting for a berth at the Port of New Orleans; and the large, post-Katrina elevated estate homes, adjacent to modest trailers, horse stables, a nature rehabilitation center, and an artist's retreat donated to Tulane University. I wonder how to account for the dizzying collisions of interests and people on the river.

5

The "Katrina Effect" and the Working Coast

Katrina's legacy to Louisiana governance today simply cannot be minimized. The winds and tidal surges of Hurricane Katrina on August 29, 2005, and fellow Category 3 hurricane, Rita, which struck the western side of the state three weeks later, not only deluged the City of New Orleans, but uprooted more than 217 square miles of coastal wetlands in its track. Industrial ports and processing facilities were underwater. Major pipelines were severed. Damage from the hurricanes burnished the state's argument that its infrastructure was both important to the national energy and shipping sectors and vulnerable. The storms disrupted 95 percent of offshore oil and gas production. Natural gas production throughout coastal Louisiana dropped by 50 percent and remained disrupted for months. Plants were damaged. Deliveries of gasoline, diesel, and jet fuel to East Coast buyers were suspended. President George W. Bush ordered the withdrawal of emergency oil supplies from the Strategic Petroleum Reserve within salt dome caverns along the Louisiana and Texas coasts. Floodwaters swamped the low-lying highway to Port Fourchon, whose once-green adjacent wetlands "resembled a vast open bay."[1]

Storm recovery efforts would require a reorganization of water and flood management and a plan to restore the beleaguered marshes. State and city leaders pitched their recovery by framing the region as a national asset with strategic importance. They leveraged Louisiana's five international deepwater ports. They leveraged the state's seafood estuaries. And they leveraged a massive oil and gas pipeline infrastructure. By disrupting the Louisiana coast, the storm had disrupted the US economy, causing fuel price spikes and shipping delays of grain and other goods to world markets.

The cause to rebuild New Orleans and the coast after the storms reenergized the stalled "America's WETLANDS" campaign.[2] "This extreme rate of loss threatens a range of key national assets and locally important communities," Louisiana governor, Kathleen Blanco, wrote after Katrina. "Pipelines, navigation channels, and fisheries as well as centuries-old human settlements and priceless ecosystems are all at risk."[3] State officials argued that "a sustainable landscape" was a prerequisite for both storm protection and "ecological restoration." They argued that hurricane protection must rely on "multiple lines of defense."[4] The pitch worked. The federal government not only approved the $14.5 billion levee wall around Greater New Orleans—which contains the largest pumping stations in the world—but it accepted the argument that the restoration of the state's coastal marshlands is an essential buffer for storm protection and oil and gas infrastructure.

Hurricanes Katrina and Rita were not only meteorological phenomena; they were also sociological events whose effects not only restructured governance in South Louisiana but also lifted a twenty-five-year ban on new oil drilling on the Outer Continental Shelf. The storms catalyzed an alignment of forces. The oil lobby that had been pushing to lift the drilling moratorium in the Gulf of Mexico found common cause with the Louisiana congressional delegation seeking a higher percentage of royalties on wells in federal waters. These two interest groups came together under the auspices of hurricane relief and a narrative of energy independence. The year 2005 was also a period of rising fuel prices and a quagmire in Iraq. The storms raised the profile and political potency of these preexisting agendas.

The Katrina Effect follows Naomi Klein's provocation of disaster capitalism through public trauma. In her book *Shock Doctrine*, Klein describes how long-held, often controversial agendas are undertaken through post-shock opportunism. The shock doctrine, which she attributes to Milton Friedman and his neoliberal "Chicago Boys" from the University of Chicago, deploys "orchestrated raids on the public sphere in the wake of catastrophic events, combined with the treatment of disasters as exciting market opportunities."[5] Efforts to support Louisiana's "working coast" after Katrina became a siren call to rebuild New Orleans and Louisiana by lifting the drilling moratorium. In the resulting debris and chaos of the storms, state officials found their long-sought federal partnership. In the following months, a multipronged political offensive was launched in the name of national energy security. The Louisiana delegation in Congress renewed efforts to increase the state's share of federal oil royalties with support from Republicans who had long advocated for more drilling in the Gulf and the Arctic National Wildlife Refuge (ANWR). Much like moves to leverage the 2022 Russian invasion of Ukraine to increase liquified natural gas (LNG) terminal permits, the oil lobby jumped at the political lubricant.

On September 5, 2005, a week after Katrina made landfall, the Republican chair of the Senate Energy and Natural Resources Committee, Pete Domenici (R-NM), said he would seek legislation authorizing oil and gas development on portions of the Outer Continental Shelf. "I'm going to go after OCS," Domenici told reporters following a hearing on gasoline.[6] A week later, on September 12, the American

Gas Association (AGA) petitioned Congress to open the eastern portion of the Gulf of Mexico to development through Lease 181. In addition, the energy lobby called for lifting the drilling ban in federally controlled Atlantic and Pacific coastal waters. The petitioners argued that the ban was implemented years ago under an energy scenario entirely different from the one facing them today. A move to drill in the ANWR was pushed into a version of a federal budget bill being debated. Environmentalists charged its supporters with exploiting the temporary energy production crisis caused by Hurricane Katrina.[7] At the same time, Louisiana's two US senators, the Republican David Vitter and the Democrat Mary Landrieu, proposed, as part of a larger hurricane relief package that was backed by Louisiana's state lawmakers, to give Gulf states a 50 percent share of the billions of dollars in federal royalties from energy companies in federal waters—and to open new areas using the same revenue formula. The money would go toward coastal restoration and flood control. The Associated Press noted, "Hurricane Katrina has reopened a national debate on energy policy, generating new congressional support for more stringent automobile fuel economy requirements, and a fresh push by the oil industry for drilling in areas now off-limits."[8] *Forbes* added, "Katrina wasn't all bad for the cause of oil and gas production. For political reasons, it may end up making Alaska and the Outer Continental Shelf more accessible."[9] On December 19, 2005, Alaska's senator Ted Stevens tied efforts to expand OCS drilling to opening the ANWR.[10] A spokesperson with the Sierra Club noted two months later that the threat to open the ANWR and the OCS was "greater than ever" that year.[11]

On September 28, 2006, twelve months after Katrina, Landrieu formally introduced a bill to boost Louisiana's royalty share for expanded OCS drilling, which was taken up by the House of Representatives. By December 2006, with the winds of Katrina at her back, her efforts paid off. Congress passed a bill cosponsored by Landrieu and Domenici, called the Gulf of Mexico Security Act (GOMESA), that increased Gulf states' share of federal royalties and opened 8.3 million acres to new oil and gas exploration in the Gulf of Mexico. Congress overrode a presidential veto by George W. Bush to do it. And Louisiana's long-sought increased revenue-sharing agreement on federal oil royalties was realized.

The move would formally enshrine deepwater oil drilling as the funding mechanism for Louisiana's coastal restoration efforts.[12] The federal government would share 37.5 percent of royalties collected on wells in the OCS with the Gulf states, with Louisiana receiving the lion's share.[13] Meanwhile, the Louisiana Legislature had been working in concert on the state level to create a legal mechanism to tie any future OCS revenue streams to coastal protection. In fall 2005, the legislature, in an extraordinary session, passed a proposed constitutional amendment, Act 69, to dedicate OCS royalties to the Coastal Protection Fund for the sole purposes of "integrated coastal protection." The amendment was ratified by Louisiana voters in November 2006, which reassured a reluctant Congress to pass the GOMESA revenue act that December. Today GOMESA, which provides up to $170 million annually, is the only major recurring revenue stream funding the state's Master Plan.[14] But its application is so expansive that it also includes improvement of infrastructure directly affected by coastal wetland loss such as elevating Highway 1—the oil highway to Port Fourchon—which was regularly deluged by high tides. The project to elevate the superhighway project to ensure that oil and gas activity will avoid disruption by rising seas was surprisingly supported by the Environmental Defense Fund.[15] But, as mentioned in chapter 4, the EDF was an original partner in joining with the oil industry to promote America's WETLAND.

SOCIAL REFORM

The Katrina Effect was also at work on New Orleans and Louisiana social policy and school reform. It unleashed a series of reforms addressing "pre-existing social problems" that had little to do with hurricane protection.[16] It illustrates how the levers of power can hide behind environmental destruction. Power, after all, is maintained by logics that seem commonsensical and are rarely questioned. "Call it the silver lining," wrote the Aspen Institute's Walter Isaacson, who was appointed by Governor Blanco to help lead state recovery efforts. "Hurricane Katrina washed away what was one of the nation's worst school systems and opened the path for energetic reformers who want to make New Orleans a laboratory of new ideas for urban schools."[17] An assortment of think tanks joined reformers and editorial boards around the country to frame the catastrophe as an exciting opportunity. Republican State Judge Joe Cannizaro called Katrina a "clean sheet" to create a "smaller safer city." Richard Baker, a Baton Rouge–area Republican congressman, noted in a speech to lobbyists, "We finally cleaned up public housing in New Orleans. We couldn't do it, but God did."[18] The school system was taken over by a state board in Baton Rouge and transformed into a complete charter system. All nine of the city's public housing projects were torn down. The city began redeveloping mixed-income housing on the same footprint, offering housing vouchers to nineteen thousand of its poorest households, whose reimbursement rates have remained stagnant as rents increased by 6 to 8 percent per year. According to the New Orleans Redevelopment Authority, most New Orleans renters—nearly three of every five—spend more than 50 percent of their income on housing, which far exceeds the national average. Four of five low-income, "cost-burdened renters" in New Orleans are African American households.[19] That was even before the pandemic and 2022 inflation levels took hold.

If anything, Hurricane Katrina provided a visual narrative of historical geographic and racial inequality in New Orleans. An examination of flood maps shows that Katrina rendered the heaviest damage to lower-lying African American neighborhoods.[20] Of course, it wasn't God that flooded them but the legacy of racial, economic, and geographic inequality through drainage politics and segregation. The Crescent City, so named for the wide crescent-like bend in the Mississippi River, had been transformed into a fortified bowl surrounded by water. Its edges were ringed by levees. Internal ridges that were built by old river meander paths prior to the levees—like Esplanade and Metairie ridges—sat a bit higher near sea level and were home to affluent neighborhoods. The city's working-class neighborhoods, most of them African American, sat at the lowest elevation—in essence at the bottom of the bowl—and regularly flood in heavy rain.[21]

By the end of the morning of August 29, 2005, there were fifty separate breaches in the regional levee system. The worst-hit neighborhoods lay in New Orleans East, flooded via the 76-mile Mississippi River Gulf Outlet (MRGO), which was dug by the Army Corps of Engineers through wetlands so that smaller vessels could avoid the yawning turns of the Mississippi. But MR-GO required regular dredging and was long criticized by environmentalists for the aggressive erosion it caused. Katrina floodwaters surged through MRGO through the backdoor of New Orleans and T-boned into the Industrial Canal at the levee of the Lower Ninth Ward, a working-class African American neighborhood where incomes averaged $16,000 a year.[22]

In fall 2005, Governor Blanco created the bipartisan Louisiana Recovery Authority (LRA) to direct post-storm recovery efforts, which more than doubled congressional appropriations for Louisiana to $28 billion.[23] Governor Blanco and LRA representatives traveled numerous times to Capitol Hill to argue for recovery funds and generate sympathetic news coverage. The *Washington Post* said in an editorial, "Louisiana is the nation's energy hub, ranking first in crude oil production and second in natural gas production. The Port of New Orleans is a major import-export route, with global merchandise exports totaling $23.5 billion in 2006. The state shouldn't have to keep begging Washington to help it rise from the most damaging natural disaster in U.S. history."[24]

Storm recovery led to a complete reorganization of water management as well. Louisiana's byzantine levee board system was consolidated into regional districts appointed by the governor, with a percentage of members required to have expertise in flood protection.[25] And in November 2005, the state legislature passed Act 8, which established the Coastal Protection and Restoration Authority (CPRA) to oversee hurricane protection and ecosystem restoration under the single mission of sustaining the land and economy of Louisiana. Act 8 stated that the loss of the state's coastal wetlands threatened its "natural, cultural, and economic resources."[26] The law articulated the economic benefits of coastal wetlands that "support recreational and commercial interests."[27] In addition, Act 8 pointed to coastal wetlands "as the first line of defense for coastal communities, including New Orleans, in the face of hurricanes and tropical storm surges." The act advocates for protection of oil and gas pipelines "through which much of our nation's energy supply flows" and gestures to the diverse coastal cultures "that have called the wetlands home for many generations." The CPRA was given oversight of all coastal activities, which had previously been located in various departments and agencies.

Act 8 elevated the CPRA as a critical player in securing federal funds in housing, environmental support, transportation, and marine and flood protection.[28] The CPRA was to implement a new "multiple lines of defense" strategy to prioritize restoration methods and projects that likewise provided flood protection. "Coastal restoration is targeted where it can provide flood protection benefit."[29] It operationalized wetland restoration to benefit certain prioritized goals. The CPRA was then tasked to produce a comprehensive master plan, which would be updated every five years. The next spring, in April 2007, a newly minted master plan was sent to the legislature. It was described as a working document with an "adaptive management framework." In her introductory letter, Governor Blanco explicitly tied the often-paradoxical effort of providing flood protection with restoring wetland ecology as a response to Hurricanes Katrina and Rita: "The death and devastation caused by hurricanes Rita and Katrina has strengthened our resolve to establish a lasting legacy of coastal protection and restoration for south Louisiana. The passage of this Master Plan is the first step in making that legacy a reality for our coastal communities today."[30]

The plan established five mission priorities. At the top was maintaining Louisiana's oil and gas industry: "Louisiana's working coast, America's Wetland, supports vital ecosystems, national energy security, a unique culture, and thousands of jobs. However, the region is changing before our eyes, threatening benefits we have relied upon for decades."[31]

THE MASTER PLAN AND THE WORKING COAST

The "2007 Comprehensive Master Plan" increased the total bill from $14 billion estimated by "Coast 2050" in 1998 to $50 billion, which is today considered an underestimate. The plan rehashed many of the arguments that state officials had been making about the value of shipping lanes, fisheries, energy infrastructure, and the seafood industry. It listed the "host of benefits" of Louisiana's coastal landscape, including protection from incoming storms by cypress swamps, barrier islands, and healthy marshes by "slowing down and reducing incoming surges of water." And it laid out the national pipeline assets at risk from coastal erosion and storm surge, which included Henry Hub, which is the pricing point for natural gas throughout North America, and Port Fourchon, which supplies hundreds of offshore drilling rigs in the Gulf. It squarely quantified the assets of the wetlands in economic terms.[32] It also plugged the coast's ecological "services" such as the North American flyway over South Louisiana, which was home to more than five million migratory waterfowl that winter in Louisiana marshes and seventeen endangered or threatened species, including the bald eagle, gulf sturgeon, Louisiana black bear, and several sea turtle species. All of this provides recreational opportunities and jobs associated with birding, hunting, fishing, and ecotourism.[33]

These lines emphasize the anthropogenic utility of protecting human settlement and economic resources.

The working coast rationalizes industrial and commercial practices that harm the fragile ecology of the area—a phenomenon akin to what the political ecologist Erik Swyngedouw calls the metabolism of an environment to extract surplus value from it.[34] The concept of the working coast that emerged just before Hurricane Katrina became the organizing rationale for the state's Master Plan for coastal restoration after Katrina to reenergize efforts to enroll federal support. As a concept, the working coast frames the state's fragile marshlands through metrics that can only be realized by continued extraction, which limits the types of interventions coastal planners consider.

The Master Plan was passed in late 2006, just over a year after the storms. It is the full embodiment of the Katrina Effect, folding Louisiana's eighty-year problem of coastal disappearance into an emergent strategy of hurricane protection. Restoration involves a multipronged approach: pumping dredged mud and sediment into marshes and onto barrier islands, securing shorelines with shoal barriers, heightening seawalls and ring levees around populated areas, and elevating homes. But its most ambitious proposal applies diversion spillways along the Mississippi River to "pulse" sediment back into the adjacent marshes. The first of ten such projects had been approved by Louisiana's CPRA by the time of this writing for an estimated $3 billion.[35] The diversions would provide a dedicated source of mud to the delta by its original progenitor, which is captured in discourses by scientists, coastal planners, and some environmentalists of returning the Mighty Mississippi to its "natural" role of land building.[36] Authors of the Master Plan say they are using "the best available science and engineering to prioritize and sequence projects for implementation."[37] But in adjudicating decisions about where and how to divvy up a limited supply of sediment, money, and other resources to protect populated areas and what authors call "critical infrastructures," the Master Plan is also deeply political.

Supporters frame it as the protector of Louisiana's working coast as well as an instrument for economic diversification for struggling coastal communities. These "political rationalities" appeal to a broad cross section of stakeholders, who may otherwise be in opposition.[38] The plan also traffics in Extractive Thinking. It establishes a future for the state's people and economy through the conditions created by the practices it supports.

The plan articulates the incumbent contradiction of living on the spectrum of survival and annihilation. "This function, combined with man-made levees and other flood control measures, have allowed Louisiana's working coast to thrive in a flood-prone area. Whether or not these citizens are able to maintain their connection to the region depends on how quickly the state can find ways to rebuild wetlands and provide adequate storm protection."[39] Therein lies the ongoing dilemma. The practices of building man-made levees "and other flood control measures" have allowed for a working coast. And this working coast is part and parcel not only of the resources that are extracted from it but also of the measures that are taken to protect it and generate it.[40]

Through Act 8 and now the Master Plan, coastal planning officials have justified saving coastal Louisiana by maintaining its industrial and economic activity that give it value around two rationales: first, by positioning the Louisiana coast as a national asset that supports national industries; and second, by raw return on direct investment in the form of economic development.[41] The CPRA argues the Master Plan will create jobs and economic spin-off effects, "to foster our state's employment capacity and contribute to the growth of Louisiana's future economy."[42] These two rationales bring the wide tableau of various interests and rationalities under a single strategy.

Five years after Katrina, despite assurances by the GOMESA supporters about better, safer drilling technology, BP's Deepwater Horizon oil well exploded, killing eleven workers and causing the world's largest oil spill over eighty-seven days. An estimated 500,000 cubic meters of crude oil gushed into the Gulf of Mexico.[43] A legal settlement against BP and its partners provided Louisiana and local coastal parishes with $6.5 billion over fifteen years. The money, which is dubbed the Resources and Ecosystems Sustainability, Tourist Opportunities, and Revived Economies of the Gulf Coast States Act, or the RESTORE Act, includes multiple civil, criminal, and punitive judgments. It is dedicated by the Natural Resource Damage Act to coastal restoration projects. The state was awarded another $2.2 billion in civil and criminal penalties,[44] which Mark Davis, director of Tulane's Institute on Water Resources Law and Policy, called analogous to "paying for a gym membership by winning pie-eating contests."[45]

By 2017, coastal plan authors had also updated a much more pessimistic estimation of sea level rise, which had flipped the 2012 Master Plan's worst-case scenario into the 2017 Master Plan's best-case scenario. Geologic surveys suggest that the rate of sea level rise is twice that estimated in the 2012 Master Plan update and may overtake the ability of the planned diversions to rebuild land.[46] State officials conceded that the subsidence of the coast could no longer be arrested but merely slowed.

Policy makers also began discussing "nonstructural" efforts for communities such as buyouts for relocation. Traveling around the coastal communities, CPRA representatives also began working with stakeholder groups to discuss several controversial ideas to encourage relocation. They include prohibiting any residential construction outside of planned levees and floodwalls, creating a buyout program for high-risk areas, phasing out the homestead exemption for property taxes in high-risk areas, requiring new commercial developments to have bonding for demolition costs at the end of their useful life or long-term vacancy, and requiring certain communities to participate in a program that lowers flood insurance rates for using flood-resistant construction. In response to the proposals, the president of Plaquemines Parish downriver from New Orleans, Amos Cormier, called it effectively condemning homes: "It's patently clear to anyone who lives here that all these proposals are against the residents' interests. Just put yourself in that position. It's the same as your home being condemned."[47]

There are also expensive updates to the federal flood insurance program, which will more than triple premiums for Louisiana rate payers, who make up 10 percent of all program participants.[48] Repeated disasters have left the state in a vulnerable position as it readies itself for active hurricane seasons, which begins each year on June 1. As the 2022 season was getting under way, FEMA recipients were still living in trailers after Hurricane Ida strafed the coastal parishes in 2021, and victims of Hurricane Laura, which struck Lake Charles in the southwestern part of the state in 2020, were still petitioning for federal help.[49] In spring 2023, lawmakers in special session passed a $45 million grant program to reward insurers for writing policies for people clinging to the state-run insurer of last resort, Louisiana Citizens, which had ballooned to over 173,000 policy holders.[50]

CONDITIONS OF POSSIBILITY

On the front wall of the large CPRA-funded river model housed at the Water Campus near downtown Baton Rouge is a quote attributed to Albert Einstein: "We cannot solve our problems with the same thinking we used when we created them." But instead of moving away from Extractive Thinking, the Master Plan allows for the continued historical practices that led to the conditions it was created under—*and* guarantees its future necessity. The plan to sustain Louisiana's working coast is inextricably tied to its extractive industries through the plan's funding mechanisms, to political rationalities that organize its logic, and to the political ecologies that render the region more vulnerable.

While tying oil royalties to mitigate damage caused to the coast may seem natural on its surface, the inverse of that logic is also true: it turns the restoration authority into an advocate for an industry that has shredded the state's wetlands and increased the danger of sea level rise. For example, in October 2017, coastal officials announced that restoration projects would have to be scaled back due to falling global petroleum prices that reduced the state's royalty check from the federal government. In response, the governor's coastal adviser, Chip Kline (later chair of the CPRA), said there was reason to be hopeful because President Donald Trump's Department of the Interior secretary, Ryan Zinke, was about to announce the largest offshore oil and gas lease sale in history: 77 million acres in the Gulf of Mexico. He said, "Zinke was here in Louisiana a couple of weeks ago, and he promised to help us move some of our much-needed coastal projects forward. He gets it."[51] More drilling places more pressure on pipeline routes through the marsh, increases the chance of accidents and leaks, and adds carbon dioxide to the atmosphere. Today we see an extension of this same logic, positioning Louisiana as a "natural fit" for storing industrial carbon below ground, thereby rationalizing the construction of new fossil fuel–powered plants.

The Master Plan, which was updated for 2023, creates the conditions for its own possibility through effects on the local ecology. It funds ring levees that protect coastal communities from flooding in the short term but whose presence disrupts the hydrological "sheeting" of sedimentation that maintains healthy estuaries. Levees not only entrap water after storms, but they encourage development in floodplains. Communities surrounded by levees are dependent on drainage pumps to remove floodwaters. Ultimately this cycle of water removal causes land within levee systems to sink. In coastal Louisiana, communities protected by levees have dipped as much as 10 feet below sea level, which leaves them more vulnerable and imminently harmed by catastrophic flooding.[52] Ecologically speaking, the vulnerability of these social geographies is reinforced by their protection, which requires subsequent intervention.

Technically, the Master Plan's multiple lines of defense strategy represents a contradiction of approaches. Roughly half of the resources in the plan are earmarked for coastal erosion and half for flood control and river dredging. And sometimes the projects to dredge the river for navigation—which exacerbates coastal erosion by disrupting the natural sedimentation hydrology of the river system—are rationalized by using the mud for wetland restoration. These are contradictory moves that undermine each other, but they are directed toward a common goal of supporting the working coast of Louisiana.

Supporters of the plan also tout its ancillary economic benefits in the form of a "water jobs cluster" that can be exported to other areas afflicted by sea level rise and environmental decline. The plan becomes its own asset: "The unprecedented investment in coastal restoration and risk reduction in the last 10 years has put Louisiana at the forefront of using science and innovation to plan a sustainable future for our coastal communities and our valuable ecosystem."[53]

The plan has become the organizing site for researchers and practitioners, scientists and design engineers, agencies, and academics focusing on moving projects "from concept to construction." The authors frame this as "a significant workforce opportunity in coastal Louisiana with employment in the water management sector projected to increase 23 percent over the next 10 years."[54]

The 2017 Master Plan cites various studies that promote positive returns on workforce investment into a water management cluster, including a Louisiana Workforce Commission report on coastal restoration spending in Louisiana that found that coastal restoration expenditures in 2010 directly created 4,880 jobs and indirectly created 4,020 jobs. Future spending estimates reported a range of total employment impact from 5,510 to 10,320 jobs annually. Total economic output of employment, including wages and "value added," ranged from $700 million to $1.3 billion. "There are two main job sectors in Louisiana that will see an increase in available job opportunities in the near future: water management and energy," according to the report. In the New Orleans region, an estimated 24,000 "job opportunities" will be created in these two job sectors by 2025, including 13,632 in water management over the next ten years, ranging from civil engineers and operations managers to analysts and construction laborers. "The 2015 Coastal Index published by the Data Center noted that within the New Orleans region, more than 9,500 water management jobs were gained from 2010 to 2014." A $25 billion investment would create 57,697 jobs over ten years and 77,453 over fifty years.[55]

Windfalls of federal and state money have changed the institutional landscape. The State of Louisiana in January 2018 opened a water campus in Baton Rouge to house the CPRA and research arm, the Water Institute of the Gulf, which issues calls for proposals and carries out its own environmental studies for CPRA projects. The sleek 35-acre Water Campus includes other tenants carrying out CPRA design, such as the LSU Center for River Studies, which operates a 90-foot-by-130-foot Mississippi River model—the largest "movable bed" model in the world—to run sediment delivery experiments. The Water Campus held a public grand opening with the media, touting that its presence has bolstered a blighted area adjacent to downtown Baton Rouge and helped elevate the city's business climate.[56] The campus is promoted in glossy brochures highlighting shared workspaces overlooking the Mississippi River.

Louisiana economic development officials tout the positive impacts that the BP legal settlement money from the Deepwater Horizon catastrophe has had on state contracts and workforce investment. "We foresee Louisiana as not only addressing its own water management issues but also developing scientific, engineering, and construction expertise in the field that can be exported worldwide," said Steve Grissom, secretary of the state's Department of Economic Development. He also emphasized the "crossover" skills from shipbuilding, maritime, and other oil and gas–related jobs. "So, the slowdown in the oil patch adds to the potential labor pool."[57] On February 21, 2018, the CPRA announced its first six winners of the local parish matching program under the RESTORE Act, which will set aside $100 million in local project funding over fifteen years. A spokesman from an industry advocacy group, Restore or Retreat, said a water management sector could help diversify the local economy from its reliance on the oil and gas industry while aiding the fight against coastal erosion: "It's been a nice silver lining to this problem that we're facing now is that this could be workforce development, it could be diversity, and it could be an economic driver for our area."[58]

Political rationality brings together disparate interests under a governing form of reason that, once it takes hold, promotes the interests of that logic. As Wendy Brown writes, "Political rationality is not an instrument of governmental practice, but rather the condition of possibility and legitimacy of its instruments, the field of normative reason from which governing is forged."[59] The Master Plan has become the normative form of reason for the benefits and opportunities it provides. It helps explain how seemingly incompatible schemes and players such

FIGURE 8. Pipeline Canal. Hurricane Ida in 2021 accelerated erosion around this pipeline canal in Lafourche Parish. The estimated 14,000 miles of canals in coastal marsh are major contributors to coastal erosion and subsidence in Louisiana. Photo courtesy of Healthy Gulf c/o Southwings.org.

as the Environmental Defense Fund and Shell Oil can join forces and serve to provide legitimacy to its logic.[60] For example, the Master Plan also enjoys the support of the powerful shipping lobby because it discursively and materially maintains the "Mighty Mississippi River" as a principal engine of commerce.[61] It rationalizes dredging the Mississippi River channel in order to pump "mud slurry" into endangered marshes.

One could think of the Master Plan as a kind of demonstration document with the wetlands as a laboratory to test speculative ideas and the rise of the water cluster sector as an industry that could be exported to other communities in an age of global warming and rising sea levels—both of which are expected to hit New Orleans particularly hard. State officials admit the publicly funded interventions will not restore the "boot" of Louisiana or many of the vulnerable communities along the coast. One might wonder, then, what the Master Plan is sustaining? Through a kind of *governmentality*, it appears that the plan is at the very least sustaining the industrial activity and assets that make the coast a viable site of investment for continued intervention. It is sustaining a rationale for intervention.

The governing logic of the Master Plan and the working coast reproduces an extractive mind-set that promotes practices that diminish the landscape to support one's livelihood. But the oil and gas industry is not the only livelihood in Louisiana. If the wetland estuaries continue to transform into open water, the state's robust seafood industry will collapse. One could argue that oil and gas development and other heavy industry is actively transforming a landscape into one that can *solely* support fossil fuel extraction. As an instrument of restoration, the Master Plan could be thought of as an *extraction machine*. It fails to call for reduction in oil and gas production, which has left thousands of miles of canals open to saltwater intrusion and "ponding" effects associated with a third of all wetland losses.[62] It contains no projects to backfill oil and gas canals, which have been identified as a low-tech solution embraced by previous restoration plans.[63] Leaving canals untouched satisfies oil interests as well as a few powerful private landowners whose access canals and wells either produce steady royalty checks or may do so again in the future with newer drilling technology or increased market prices.[64]

An estimated 80 percent of coastal land in Louisiana is privately held, most of it by a handful of large landowners residing outside of Louisiana. Conoco, for example, owns 700,000 acres.[65] Randy Moertle, who represents a consortium of six South Louisiana landowners that collectively own 185,000 acres and sit on several stakeholder coalition boards, including America's WETLAND and Ducks Unlimited, said that backfilling canals is extremely unpopular among his cohort. Moertle's consortium typically lease their mineral rights to oil and gas companies and use their surface rights for alligator hatchlings, ranchland pasturing, duck hunting, fishing, and other revenue-producing outdoor activities. What irks them, according to Moertle, is when a scientist will propose a marsh restoration project on their property without collaborating with the landowner. "They might say, 'let's put a marsh here,' but that's on top of my alligator hatchlings. That's not going to happen." For all intents and purposes, without the landowner's consent, any effort to backfill canals would require eminent domain and a legal "taking" by the state and end up in court litigation, which will take time and resources away from the unfolding catastrophe of coastal erosion.[66] Backfilling is also unpopular with fisherman, said Jim Tripp of the EDF, who characterized backfilling canals as "buying Peter to pay Paul" because the sediment would have to come from somewhere. The lack of sediment is an ongoing constraint cited by coastal planners. Even river sediment—if directed into the marsh—contains about half the volume it once did because of urban hardscape development during the twentieth century throughout the Mississippi River basin.

Backfilling canals is too individualized to be considered part of the large-scale, system-wide approach that the Master Plan takes, according to Denise Reed, former science director of the Water Institute.[67] Creating a backfill program would require a large mobilization effort to directly siphon mud and small amounts of material to different places, she said. Meanwhile, one of the early coastal restoration advocates, Mark Davis, says that the longer backfilling is neglected, the less sediment is available for it. When the state first considered backfilling in the 1980s, the spoil bank ridges of mud cuttings along the sides of canals could have been pushed back into the water channel and prevented subsequent saltwater intrusion while providing platforms for vegetative growth. Those solutions were actively fought by the oil and gas lobby and screened out of the Master Plan. Today many of the banks themselves have compacted into the eroding conditions they helped cause through hydrological disruption.[68] Their neglect has been productive for the political interests that have long resisted them.

As an extraction machine, the Master Plan also fails to build on findings by USGS surveys on subsidence hotspots in the marsh. These spots correlate to periods of rapid removal of crude oil that may have been caused by either depressurized well cavities beneath the surface or deep well brine that may have triggered subterranean fault activity.[69] There is no public discussion by coastal planners to repressurize old wells with fluid to halt subsidence as is required in California and other places.[70] Instead, the Master Plan focuses on implementing system-wide projects like diversions, which have been met with resistance by many coastal communities whose residents rely on the brackish estuaries for seafood harvesting, oyster farming, and fishing.

TECHNICAL DISAGREEMENTS

Opponents of diversions argue that they are unpredictable, slow, and expensive. Two pilot diversion projects created in the 1990s by the Coastal Wetlands Conservation Grant Program, or Breaux Act, have produced mixed results.[71] Public forums held by the CPRA in coastal communities are often punctuated with heated discussions and acrimony. Social scientists generally have argued that the social effects of the Master Plan and diversions need to be considered with the same priority as the technical efficacy of land restoration. Good science is essential, but because environmental management is fundamentally a human activity, effective predictions of human impacts demand, at the very least, paying equal attention to the social, political, cultural, and economic systems in which environmental management takes place.[72]

In addition to the technical challenges, the projects are controversial due to expected impacts on downstream communities and commercial fisheries. They are opposed by an assortment of interests, including residents who may be forced to relocate, commercial fishing captains, oystermen, and other stakeholders in the commercial seafood industry worried about river infusions in the saltwater ecology. The state's lieutenant governor, Billy Nungesser, launched a political offensive in 2021 around southern Louisiana as he spoke against diversions. There are also environmental concerns: mortality rates of dolphins, Kemp's ridley sea turtles, and other wildlife that tend to suffer when the Army Corps of Engineers opens the Bonnet Carré Spillway into Lake Pontchartrain during flood stages of the river. In addition, questions persist about the resilience of shallow Louisiana marshes to pollutants in the Mississippi River that currently pour into the Gulf of Mexico, such as pesticide runoff from midwestern farms and plastic litter.

The state's powerful oyster lobby is also against the diversions, which they fear will "over-fresh" their leases. As recently as January 2023, the chair of the Louisiana Oyster Task Force said the $3 billion Mid-Barataria Sediment Diversion would

FIGURE 9. Shrimp boats along Bayou Terrebonne at the Indigenous community of Point-au-Chien. Louisiana shrimp catches have dropped by more than half, to 74 million pounds from 2000 to 2021. Local shrimpers blame cheaper imports. Many shrimpers also work offshore on oil rigs. Photo courtesy of Kerry Maloney.

devastate the seafood industry and put the oystermen out of business permanently. "Oysters will become extinct due to this Mid-Barataria Sediment Diversion, and it said that in the [Corps] economic impact statement as well," said Mitch Jurisich.[73] In 1994, the plaintiffs representing the state's fifteen hundred oyster leases won a federal class action lawsuit over damages from the Caernarvon freshwater diversion, which was built in 1990. They wanted one hundred years of revenue from the leases and were initially awarded a $1 billion jury settlement. However, on appeal the Louisiana Supreme Court reversed the decision and found that the state was not liable because of a "hold harmless" clause in the contract.[74] The presiding judge reduced their harm to three years of revenue with a fair relocation fee.

However, the Caernarvon diversion led to the state's oyster lease moratorium in 2002. By then, more than 400,000 acres of state-owned water bottoms had been leased for oyster production, which raised more than $1.2 million in annual revenues. In 2016, legislation established a framework to lift the moratorium.[75] State officials are sensitive to the various positions and pots of funding sources related to projects, particularly the proposed sediment diversions. In a 2018 guest newspaper column, CPRA director, John Bradberry, took pains to specify that the two sediment diversion projects that are advancing through federal permitting, the Mid-Barataria and Mid-Breton Sediment Diversions, are being funded by "money available from the criminal settlement of the 2010 Deepwater Horizon oil spill—not your tax dollars."[76] An outspoken fishing boat captain, George Ricks, who represents one such group, Save Our Coast, questions why the state is embarking on unproven projects that cost billions of dollars when they could more quickly pump dredged slurry into marshes, which would have an immediate effect, versus waiting years for the marsh to recover if at all. "We need land now," says Ricks, who runs charters out of St. Bernard Parish downriver of New Orleans. "The only way we can do that is by dredging." At a meeting one afternoon at a family-owned diner in St. Bernard Parish, the fishing captain said that it's obvious to him why the state favors big diversion projects over small, targeted efforts. "Look at all the money behind this," he said. Tanned and with a blackened mustache, Ricks repeats a familiar argument of many others dependent on the seafood industry. But he is more concerned about saltwater catches being replaced by bass and other freshwater fish. "What makes Louisiana so unique," he said, "is the saltwater fish in the marshes."[77] In the meantime the marshes themselves are getting saltier as the estuaries erode. The sighting of bottlenose dolphins just a stone's throw from the Pointe Aux Chenes Marina at the water's edge in Terrebonne Parish would have been quite a spectacle a generation ago. Today they are common. Yet Ricks and others feel that the displacement effects of the old pilot diversion at Caernarvon, which operates at 8,000 cubic feet per second (cfs), will be dwarfed by the much larger sediment diversions the state is planning.

When the Bonnet Carré Spillway gates were opened for a record forty-three days in 2019, the freshwater caused large marine die-offs and algae blooms in brackish Lake Pontchartrain and the Gulf Coast. The Institute for Marine Mammal Studies in Gulfport, Mississippi, reported that twenty-five dead Kemp's ridley sea turtles, which are endangered, and ninety-three dead dolphins washed ashore in the first six months of the year, triple the average number of strandings. Mississippi's Department of Environmental Quality closed all twenty-one state beaches in response to toxic algae blooms. Louisiana seafood harvests also suffered. Louisiana governor John Bel Edwards requested a disaster declaration on shrimp, oysters, and other fisheries from the US Department of Commerce.[78] The impact served as a harbinger for worried fishermen. "What we're seeing is a preview of what's going to happen," Ricks told the board of the CPRA at a meeting in June 2019. "Is this what we want to do?"[79]

Others are concerned about pollutants from the Mississippi River. While there is evidence supporting the efficacy of marshes to filter municipal effluence, it is not at all clear if Louisiana's degraded marshes can filter what's flowing down the Mississippi River.[80] Currently, farm pesticides and nutrient runoff at the river's mouth generates a hypoxia dead zone of algae whose plume rivals the size of Vermont and consumes enough oxygen to suffocate marine life.[81] The openings of the Bonnet Carré Spillway in 2019 to relieve flood levels from the Mississippi River into Lake Pontchartrain and the Gulf of Mexico contributed to the single largest hypoxia dead zone in the Gulf of Mexico. In an example of resistance, the local government of Plaquemines Parish tried to withhold permission in 2018 for the state to take soil samples for its multibillion-dollar diversion structure that could send as much as 75,000 cubic feet per second of sediment-laden freshwater from the Mississippi into brackish Barataria Bay. In response, the state threatened to withhold other restoration projects until the local government complied with its requests.[82]

Beyond that, some communities in the path of diversions will be forced to move because of increased water levels.[83] Located 25 miles south of New Orleans, the town of Jean Lafitte discovered that in the 2012 version of the Master Plan, it was not on the list to be included in the new "Morganza-to-the-Gulf" levee system. It was also in the path of the proposed Mid-Barataria Diversion. The town's leadership set out to increase the community's strategic value by securing so much public infrastructure that it would become too valuable to abandon. As profiled in the *New York Times*, the town's longtime mayor, Tim Kerner, had successfully secured a suite of state projects, including a 1,300-seat auditorium, a library, a wetlands museum, a civic center, and a baseball park. "Jean Lafitte did not have a stop light, but it had a senior center, a medical clinic, an art gallery, a boxing club, a nature trail, and a visitor center where animatronic puppets acted out the story of its privateer namesake." It mattered less how much the facilities had been used but that they existed. "Do we lose that investment, or do we protect it? I hope people will see that, hey, not only are we fighting hard to exist, but, you know, maybe this place is worth saving," Kerner said.[84] He was able to convince state planners to establish limited levee protection in the 2017 Master Plan update.

On my drive through Jean Lafitte in 2019, I saw dozens and dozens of brick ranch houses now standing atop 20-foot pilings. Hurricane Ida punished the area in 2021, leaving a layer of thick black mud over most of Lafitte's streets and yards. The smell of stagnant swamp hung in the air. The National Guard had to install a temporary bridge to carry cars across Bayou Barataria. Mayor Kerner was lobbying again for heightened federal levee protection on social media, arguing that it's costing more money to rebuild than to protect.

Jean Lafitte may have simply taken a page from New Orleans. Rather than retreat after Katrina, city leaders doubled down. The city's airport authority opened a $1.3 billion airport just before the COVID-19 shutdown. The state's congressional delegation has been deploying the same argument for the value of the coast ever since it failed to win passage of the Conservation and Reinvestment Act, discussed in chapter 4. The state's way forward is through ambition for a future, and, it seems, the working coast gives them an effective strategy.

Meanwhile, a contingent of researchers argues that building sediment diversions without addressing the thousands of miles of oil and gas pipeline canals throughout the coast may increase subsidence.[85] Some ecologists and other marine researchers also criticize the diversions as being the wrong tool for marsh restoration.

The ecologist Eugene Turner argues that the thick, gnarly cord grass is needed to keep the dirt together. It's not just mud, but organic plant matter that is needed to create marsh that is sturdier than the marsh created by the Caernarvon pilot diversion site. The Caernarvon marsh is "floating mat," says Turner. It was pushed up like an accordion during Hurricane Katrina because it is infused with water loaded with nitrates and other fertilizers that are harmful to marshlands when overloaded. "Flooding them with Mississippi water is the worst thing you can do. The state built a geology model when this is living organic marsh, a biological system."[86] The rebuilt marsh by Caernarvon has shallow roots. All you have to do is reach down and grab a tuft. "They come out in your hand," observed Ricks.[87] His concerns echo the results of a 2011 paper, "Freshwater River Diversions for Marsh Restoration in Louisiana." It analyzed satellite images of the areas of three freshwater diversions and found that, through 2009, marsh area had not grown significantly at the diversion sites. The research also found that the diversion regions suffered more damage during Hurricane Katrina than other areas, apparently due to freshwater plants being more fragile than brackish-water plants.[88] The authors concluded that the scientific basis for river diversions needs to be more convincing before embarking on a strategy that may result in marshes less able to survive hurricanes. Supporters of the diversions outlined in the Master Plan note that its diversions are modeled to direct sediment into marshes, unlike the Caernarvon freshwater diversion.

THE GREENWASHING EFFECT

As an individual case, Louisiana reflects the larger social and environmental impact of twenty-first-century energy policy. It has fostered a plan that deploys science for coastal restoration efforts that ends up rationalizing the state's petro-economy. The common sense that it relies on reflects a global logic reproduced through international oil and gas production networks where oil companies either extract without hindrance or buy what Toby Miller describes as "social licenses to operate."[89] By purchasing other goodwill offsets, oil companies produce a "greenwashing effect," which is particularly insidious in Louisiana, which is both one of the nation's largest producers of fossil fuels and singularly vulnerable to sea level rise.[90] Louisiana's coast constitutes 40 percent of the US coastal marshes and 80 percent of its losses.[91] Greenwashing allows corporations to act as good stewards even though their primary concern is extracting profit for shareholders at minimal costs. In greenwashing campaigns, corporations routinely describe themselves as citizens while principally pursuing economic interests. As Miller writes in *Greenwashing Culture*, "Their restless quest for profit unfettered by regulation is twinned with a desire for moral legitimacy and free advertising—based on 'doing right' in a very public way, while growing rich in a very private one."[92]

I once attended a meeting of the America's WETLAND coalition composed of high-profile environmentalists, landowners, and restoration planners at Nicholls State University, 50 miles southeast of New Orleans at the steps of Terrebonne Parish's receding coast. The meeting was sponsored by the international mining and petroleum company, BHP (formerly BHP Billiton). A spokeswoman for the Australian-based multinational said it intended to operate in the Gulf of Mexico for decades to come: "Part of who we are is sustainability and partnering. We want to make sure that we are part of a stewardship to leave things in a better position than when we arrived."[93] BHP in February 2017 invested $2.2 billion in the new Thunder Horse water injection platform owned by BP, which marked BP's first project in the Gulf of Mexico since its 2010 Deepwater Horizon oil disaster (still the world's worst environmental disaster on record).[94] At the WETLAND meeting, Rachel Archer, who is BHP's general manager for Gulf of Mexico operations, stressed their commitment to social responsibility: "We need to be able to demonstrate we are responsible, be good stewards."[95] She pointed to the company's international presence as a point of its stewardship, saying, "We are global mining and petroleum—beneficiaries of these resources all over world. That comes with a social responsibility." The comments were remarkable for a company in the midst of settling a $51 billion compensation claim for the massive mining dam collapse in 2015—called the worst environmental disaster in Brazil's history—that killed nineteen people, destroyed three towns, and contaminated 280 miles of river with iron waste.[96] In June 2018, the company issued a report that five of its mining dams in Brazil and Australia "are at extreme risk of collapse," which would cause damage and loss of life.[97] BHP was also fined $25 million by the US Securities and Exchange Commission (SEC) in 2015 related to its "hospitality program" at the 2008 Summer Olympics in Beijing that provided officials of 176 government- and state-owned enterprises an all-expenses-paid package to attend the Games. The SEC found the company violated the Foreign Corrupt Practices Act by inviting officials from at least four countries where BHP had interests in influencing the officials' decisions.

It is challenging to be a good local steward when the profit centers and headquarters of companies are thousands of miles away, says Mark Davis, which adds to the conundrum in Louisiana, which is full of "middle managers."[98] Under the current legal architecture, oil-producing landowners are simply incentivized to turn areas into what Julie Maldonado calls "energy sacrifice zones." In sacrifice zones, human lives are valued less than the natural resources extracted from a place.[99] Such extraction activity generates shareholder profits and state tax revenues. Rapid resource extraction that denudes the surrounding area makes those within the sacrifice zone increasingly vulnerable and marginalized, causing further economic or physical displacement. To understand environmental degradation and displacement as what Maldonado calls "tacit persecution," we must understand how such conditions are created.

In Louisiana, sustainability signifies sustaining a healthy business environment at the expense of other ecologies of social and environmental health. Discourses of *sustainability* flow throughout the state's Master Plan and discussions of the working coast.[100] Active measures to intervene and reverse coastal erosion are undertaken to continue extraction. So, ultimately, the cycle continues. It is part of a continuum of Louisiana's political economy of extracting and exhausting its natural resources—from old growth timber and muskrat fur to fisheries to oil and gas. Extracting resources from the land is part of the state's identity, which provides the cultural cover of continued extraction. Through an effective greenwashing campaign, the industrial polluters and oil companies that have operated for years in the Louisiana wetlands and the Gulf of Mexico joined forces with environmentalists in the early 2000s to successfully underwrite a national campaign that framed the oil industry not as the cause of land loss but as one of its victims. We can link the deployment of the working coast precisely to this campaign, which was developed after efforts failed in 2000 to win federal support for the state's first comprehensive restoration plan, "Coast 2050." This argument for national relevancy of the coast's industries ties the preservation of the coast to the very practices causing coastal erosion. It cements the uneasy mixture of oil and water that is part of the ontological dilemma of Louisiana that I explore in the introduction.

As groundbreaking as "Coast 2050" was in terms of its strategic and regional approach, coastal planners were unable to attract a federal partner without a financial calculus that dollars invested in coastal restoration would be justified by financial return. They had to quantify the value of the wetlands through its industrial productivity, which continues to limit imagined futures for the land. Today, as part of any restoration argument, coastal advocates and industrial interests highlight the industrial productivity of the coast to justify financial returns on investment. The moniker "working coast" as the state's linchpin issue provided talking points for saving the coast while underplaying the problematic strategy of protecting an industry from the destruction it causes. This action tacitly shifts the financial burden of restoration onto the federal government and US taxpayers.

While the plan traffics in discourses of "resilience" on behalf of some communities, it brackets off expectations of sacrifices by capital interests. Without curtailing drilling, the state leveraged the devastation wrought by Katrina and Rita to successfully renegotiate the state/federal royalty share through GOMESA, something they had failed to accomplish in 2000. They were able to build on a legacy of Extractive Thinking by using Katrina and coastal erosion as a vehicle to intensify oil and gas drilling—which would presumably fund a master plan to mitigate damages from oil and gas drilling. This required some admission of the industry's historical destruction in the state's wetlands, which they could ostensibly mitigate by further energy production. They aimed their ire instead at the Army Corps of Engineers' leveeing of the Mississippi River, which is only one of several causes of coastal erosion and subsidence. As a result, they produced a plan that I argue rationalizes further activity and reproduces a need for itself and future mitigation measures. In this way, the plan reproduces the conditions for its own possibility. It bolsters an economy that requires further interventions in the landscape, whose disappearance exposes more people and pipelines to escalating storms.

All of this begs a fundamental question of whether Louisiana can be separated from the economic rationalities that set the crisis in motion and continue to justify an unending continuum of intervention. Simply, can Louisiana and New Orleans exist without a working coast that appears to be both sinking it and rationalizing a plan to maintain it? Can we envision the existence of New Orleans or Louisiana without its accompanying dependence on extraction, that includes not only deep-draft shipping along the Mississippi River and a robust oil, gas, and petrochemical industry but also measures to mitigate its damage funded by the extraction itself? Or is Louisiana simply fated to become the nation's *disaster laboratory*, either a cautionary tale or a model of resiliency for other governments in the crosshairs of the approaching onslaught caused by global climate change?

A Modest Proposal

When Katrina's corkscrew reached the tip of Louisiana's boot in 2005, I was safely ensconced in Baton Rouge 85 miles away from New Orleans with my fiancée, who had evacuated the city the day before. Our electricity failed just as an oak tree branch broke through the living room window. After the wind had safely died down that morning, I checked into work at the Advocate *newspaper and drove around town taking inventory of uprooted trees and broken windows. News was not yet trickling in about the devastation to the south. Baton Rouge had survived the storm with some relatively minor damage and an extended power outage. I was really preoccupied by our engagement the day before. I had proposed at a farmhouse north of Baton Rouge of my friend Prentiss, who was out of town. During our first glass of champagne, Prentiss's in-laws came up the drive. They had evacuated from New Orleans, bringing with them freshly caught fish and homemade mayonnaise. We sat around the kitchen table eating fried fish and discussing wedding plans in what was the last stretch of levity before an enduring shadow. In the following days, Jessica would discover the loss of her apartment, car, and most of her belongings. She spent her remaining year of medical school on the move, volunteering at medical shelters around Baton Rouge, and she finished the year at Baylor University in Houston, which had taken on students from her class. I filed the last of my Katrina stories before leaving for San Diego in 2006 for Jessica's residency in the navy. As a Louisianan, I spent the formative period of my life in New Orleans, where I attended college, met my fiancée, and started my career at the Associated Press and Mayor's Office. For the next twelve years, I watched from afar the uneven recovery. I missed Louisiana, terribly at times. New Orleans at one point sat almost completely empty of people. Each had to decide on their own whether to return. Post-Katrina recovery was an invitation to explore my own politics of belonging.*

We were lucky. We were out of the city before the ordered evacuation. We had a place to stay. We were not marked by race. No coerced busing, dallying governors, or nervous Samaritans. No guarded bridges or barricades. In the days following, the

Baton Rouge Convention Center became a shelter for evacuees. The LSU basketball arena was a makeshift hospital where Fats Domino was discovered by his niece and put up in the LSU dorm room of the football team's quarterback. The most destitute were plucked from rooftops and picked up in volunteer ski boats and makeshift rafts, then dropped on patches of dry land from where they trudged to the gathering masses at the Superdome and the Convention Center and on highway overpasses. Amid the chaos, there was also a familiar specter of the black threat of fear-based rumors that had spread—discourses of looting, rapes, and violence. Rescue boats had SWAT teams with rifles hoisted ready. Governor Kathleen Blanco had announced that National Guardsmen had returned from Iraq and were there to prevent looting. "They have M16s, and they're locked and loaded," she said. "These troops know how to shoot and kill, and they are more than willing to do so if necessary, and I expect they will."[1]

Evacuees bused and flown to cities throughout the country in a massive dispersal of poor people were treated, if wearily, as visiting guests that would hopefully be on their way. Some ended up staying where they landed. Others moved on. Ten years after the storm, there were at least 100,000 fewer people in New Orleans, most of them African American. Where did they go? Lolis Eric Elie, a writer and former columnist for the Times-Picayune, *wrote a piece for the tenth anniversary of Katrina to try to answer a question that only a few people seemed to be asking: Why come back? "Some of us came back because we had a cousin or auntie, and they said we could stay by them until we got it figured out," he wrote. "Some of us came back because we knew they didn't want us back."*

I met an evacuee in 2007 while I was reporting on wildfires outside of San Diego. He had been bused from New Orleans to Nevada, where he ran into trouble, got arrested, and ended up working in a prison fire brigade program to support western US fire suppression. "You from New Orleans?" he asked. "You were in the dome?" I was not in the dome. I was lucky . . .

"Some of us came back because Richard Baker, the Baton Rouge congressman, was right: 'We finally cleaned up public housing in New Orleans. We couldn't do it, but God did.' Some of us came back to fight for our homes in the Lafitte, in the Magnolia, in the B. W. Cooper, in the Melpomene, in those timeworn fortresses, those unflooded, moldless bricks."[2]

Katrina-related flooding between New Orleans and Baton Rouge disturbed five hundred industrial facilities and five Superfund sites. Six major oil spills occurred, and seven underground oil storage tanks were disturbed.[3] Floodwaters rushed into the city until they rose to the equilibrium of sea level. Waterlines stayed fixed at attic levels, over cars, at the front doorbell, for over a week until temporary repairs could be made to the levee system and the water could be pumped out again.

Three weeks later, Hurricane Rita collided with the western half of the state and reflooded some of the loosely patched levees. By then, the city's borders were manned by sandbags and National Guardsmen. When we crossed through on my press credentials two weeks after Katrina to salvage Jessica's belongings, we opened the door of her duplex to a hot stench of mold. Her toilet and sink were stained with dark brown film. Everything had floated and swirled around the room. A chair was wedged waist-high into the bedroom doorway. The grass up and down the street was dead. There were no birds in the trees. The water level was above the roof of her car. As we attempted to lift and remove the waterlogged futon mattress in the front living room, I dropped the mattress, walked outside, and started to dry heave.

6

Mud, Plastics, and Cancer Alley

As Extractive Thinking through oil and gas production became enshrined in the strategy to restore Louisiana's wetlands, the state was doubling down on attracting investments in petrochemicals, liquified natural gas terminals, and pipelines to supply a growing presence of industrial facilities along the Mississippi River corridor, once known as the German Coast. In a matter of decades, Louisiana has become ground zero for the production of plastic "nurdle" pellets that are shipped around the world to be molded into plastic products. Nurdles are loosely regulated and often spill onto docks and riverbanks during shipment. In August 2020, a containership hitched to the Napoleon Avenue Wharf at the Port of New Orleans dropped a 25-ton container holding hundreds of millions of plastic pellets into the Mississippi River. A thunderstorm had pitched up the already swelled river. Rapids broke the massive cargo ship free of its moorings and sent the container into the river. Authorities waited three days for the storm to settle before attempting to retrieve it. As it was being lifted by crane from the water, the boxcar-like container door opened and spilled 750 million pellets into the water. A local environmental scientist declared it "a nurdle apocalypse."[1] Over the next month, regulators debated who was responsible for cleaning up the accident as scores of tiny pellets washed up along the sandy banks of the river. Eventually, the ship's owner, CMA CGM, hired crews to use leaf blowers and butterfly nets in a halfhearted attempt that ultimately fell on volunteers to painstakingly collect tiny plastic pellets. The pellets bind with pesticides and pollutants and are eaten by birds, fish, and other wildlife, entering the human food web.

Such catastrophic events may occupy news outlets for a time. But the general supply chain is rife with smaller, unmanaged spills of the lentil-size pellets that have become part of an urgent crisis in ocean litter from New Orleans to Sri Lanka. Though not far afield from liquid hazardous waste like oil spills, which are managed by the US Coast Guard, plastic pellets are largely unregulated.

They are ground down by the relentless pounding of oceanic waves into microplastics ranging in size from a grain of rice to something smaller than a dust particle.

Extractive Thinking has transmogrified delta mud into a landscape of plastic waste, whose production creates terrible health conditions for residents who are sacrificed as if they were mud. Here, I examine the ecology and inherently racialized harm behind plastic production as the antithesis of mud and petrochemical production built on the legacy of the plantation economy. Thousands of miles of pipelines carry billions of barrels of oil and trillions of cubic feet of natural gas (produced onshore and offshore) across Louisiana's tidal marshlands through an intricate network of pipelines to refineries and petrochemical plants dotted along the 85-mile Mississippi River industrial corridor, where inland fence line communities face direct industrial exposure. The writer John McPhee noted that Louisiana's industrial plants on the Mississippi "made the river glow like a worm."[2] While the petrochemical corridor is touted by advocates as an important source of thousands of manufacturing jobs, its other monikers, Cancer Alley and Death Alley, speak to the health outcomes and environmental racism that befall fence line communities, which are exposed to the highest concentrations of chemicals in the country, linked to cancers and other respiratory and prenatal illnesses. The prodigious production of oil, natural gas, plastics, resins, fertilizer, and LNG keeps the pipelines humming twenty-four hours a day and showers toxic emissions onto predominantly African American communities. That's in addition to the threats of routine accidents and the constant noise of industrial operations.

Entering the area from New Orleans, the horizon appears incongruent: a vista of tall—sometimes spewing—smokestacks rising over a thicket of cypress forests. Along Highway 61, a dystopian landscape sprouts out of the forests and swamps punctuated by cylinder depots, winding pipes carrying streams of fluid between processing units, distillation columns, enormous storage tanks, and rusty stacks. Along the road, the metallic landscape becomes indiscriminate sprawl as one plant begets the next, differentiated only by branded signs at gated entrances. Elevated pipes cross over the highway to storage tank yards. Steve Lerner describes the scene: "There are catalytic cracking towers, stacks topped by flares burning off excess gas, huge oil and gasoline storage tanks, giant processing units where oil and its derivatives are turned into a wide variety of useful chemicals, and a Rube Goldberg maze of oversized pipes."[3] Here is the stark underbelly of capitalism and consumption, "part of the front end of the system that has forged the American lifestyle by making products cheap and convenient," writes Lerner.[4] Some stacks, blackened and browned, offer a vivid contrast to promotional materials of squeaky-clean facilities disseminated by the Louisiana Chemical Association.

Today, there are two hundred industrial chemical plants and refineries between Baton Rouge and New Orleans. This massive "oil assemblage" of plants produces everything from insecticides and fertilizers to jet fuel and neoprene rubber. The corridor hosts the world's largest manufacturer of Styrofoam.[5] Louisiana is the second highest producer of US petrochemicals in an area much more condensed than the top producer, Texas. Viewing this industrial landscape up close is depressingly oppressive. Train whistles and clanking railcars echo over the landscape. Pungent odors waft here and there. It's hard to imagine living within the walls of a modest bungalow amid mammoth storage tanks and hissing pipes.

When President Joe Biden rolled out his $1.2 trillion infrastructure proposal in 2021, he highlighted issues of environmental justice, which some authorities in Louisiana rebuked. "With this executive order, environmental justice will be at the center of all we do addressing the disproportionate health and environmental and economic impacts on communities of color—so-called fence line communities—especially . . . the hard-hit areas like Cancer Alley in Louisiana or the Route 9 corridor in the state of Delaware," Biden said. Louisiana's Republican senator, Bill Cassidy, shot back that the president's use of the term "Cancer Alley" was insulting. "I'm not going to accept that sort of slam upon our state. It sounds like great rhetoric. But again, I don't accept that slam."[6] There are plenty of like-minded industry advocates and lobbyists whose rebuttals are aimed to cast doubt on links between air pollution and poor health.

SOWING DOUBT

Whether it's doubt about climate change or the direct causal link between particulate emissions and cancer, *doubt* is an effective defense that slips into the nooks and crannies of correlative evidence. State regulators often downplay health claims by citing a lack of direct evidence when issuing industrial permits. Or they point to the Louisiana Tumor Registry that shows nominal elevations of documented cancer rates. Critics say the Tumor Registry casts too wide a geographic net and fails to account for higher incidences of cancer closer to plants. The registry also fails to document the constellation of illnesses other than cancer, such as respiratory diseases, skin irritations, mental impacts, and miscarriages. It fails to collect data on contributing factors or environmental conditions that people with cancer have been exposed to. "The Tumor Registry doesn't measure exposure to chloroprene or any other chemical," said Kimberly Terrell, who has a PhD in conservation biology and is a staff scientist with the Tulane Environmental Law Clinic. "They measure cancer, which is only part of the equation."[7] This leaves sickened neighbors to offer anecdotal accounts to rebut official estimates. "Proof of causation in the case of cancer or any other suspected environmentally related disease is difficult to produce."[8] Such studies need to not only analyze those living in the proximity of plants but also adjust for those who also drink, smoke, or have a genetic disposition to cancer. Length and concentration of exposure would have to be factored in, in addition to many other factors. "The beneficiary of this inaction is the chemical industry, which can unequivocally state that there is no proof that their pollution harms neighboring residents."[9] Oil and gas and petrochemical industries are also often awarded long-term local property tax breaks that starve local communities of important revenue sources for public services.[10] Legislative bills are regularly passed to shield industry operators from regulatory oversight and culpability.

GLOBAL WARMING AND A SECONDARY MARKET

As global warming depresses the industry's image and bolsters more sustainable energy alternatives, the oil and gas industry is reportedly shoring up product demand through downstream, value-added production in petrochemicals.[11] Fossil fuel manufacturers see plastics as the next frontier of market demand as vehicle consumer demand shifts to renewables. Plastics offer the fossil fuel industry a second life for polluting carbon-based products.[12] Here in Louisiana, billions of dollars in new plastic manufacturing facilities are being planned and at least eleven new LNG terminal projects have been approved, fed by the very same pipelines shredding the state's wetlands. Cumulatively, the new projects would increase Louisiana's emissions by 38 percent.[13] All of this means an onslaught of plastic production, litter, and continued fossil fuel extraction and new pipeline infrastructure through coastal wetlands.

In general terms, petrochemicals are chemicals derived from "substances or materials manufactured from a component of crude oil or natural gas."[14] Starting in the 1950s, oil refineries began to "mine" their process streams for compounds to make higher-value products. Shell in Emeryville, California, and Standard Oil of New Jersey began studying derivatives of their raw materials. By the mid-1950s, American Cyanamid had expanded its operations into Louisiana, followed by Monsanto, to produce fertilizer and ammonia.[15] From 1955 to 1956, approximately $600 million were invested in new and expanding petrochemical plants in Louisiana.[16] "From 1964 to 1968 . . . petrochemical growth in Louisiana outpaced all other states, including Texas."[17] Companies like Union Carbide and Dow Chemical relocated here. Refineries such as Shell Petroleum Company generated chemical sister plants for secondary markets for newly discovered products like antifreeze, tires, plastic food containers, trash bags, and laundry detergent.[18] In the late 1960s, demand for fertilizers surged. Louisiana became a favorite spot for new ammonia plants, again predicated on the availability of cheap oil and natural gas feedstock.[19] The plastic boom followed, with dozens of plants manufacturing polyvinyl chloride (PVC), polypropylene, synthetic rubber, polystyrene, melamine crystal, and isocyanates for urethanes.[20] "The petrochemical plants have, to a large extent, located where they may obtain the refined off gases which formerly were burned as fuel or flare gas."[21] Since a

large portion of petrochemical raw materials are by-products of refining operations, "it would be expected, then, that petrochemical plants would be most
common where the greatest concentration of refineries are found."[22]

Plastic is perhaps modernity's ultimate ruse. A material seemingly impervious
to leakage, it promises to hold contaminants in a hygienic embrace. Its appeal lies
in its sanitary separation, as well as its cheapness. Most plastics are made from
extra feedstocks of gas and carbon-based material that are produced through the
refining process. Monetizing such waste products that began in the second half of
the twentieth century appealed to everyone from housewives to commercial car
makers. Today, plastic is the most ubiquitous product in the world and the hard-
est to dispose of. And even its value proposition of keeping things separate and
preserved is false. Over time, plastic itself degrades.

Microplastics have become ubiquitous in not only our visible environment but
also the micro space. Plastics have penetrated our very cellular walls. They are
in our bloodstream and our food webs. Microplastics are found on the remotest
mountaintops, borne by airstreams. They permeate the air of urban cities from
vehicle road dust. Researchers have found plastics entering the atmosphere
from sea spray. And plastic production over the next decades is expected to rise.
There is no place on Earth or within Earth that plastic has not penetrated. "It is
in the Arctic, the Mariana Trench—the deepest place on earth, over ten thousand
meters beneath the surface of the Pacific Ocean—and on remote mountaintops in
the high altitudes of the Pyrenees."[23] Despite the fact that plastic was designed as
a barrier, it has become part of contemporary nature. "Its chemical by-products
have been found in everyone who has been tested. The world is now plastic."[24]
Indeed, it seems as if plastic (like the river) has *crevassed* from its intended location to infiltrate everything. There is no natural/synthetic divide anymore.

The passage of the 1966 Federal Water Quality Act would ironically place
more pressure on Louisiana communities. Companies began factoring pollution
into their costs. Rivers with a high discharge rate, like the Mississippi, made
an ideal location in light of coming regulations.[25] As chemical plants and refineries were pushed away or discouraged from expanding elsewhere, they were
welcomed with open arms in Louisiana, where they began dotting the landscape of former plantations.[26] Many plants bought riverfront property from
former plantation owners who then moved, leaving poorer and minority neighbors behind.[27] "Enabled by state zoning, a wave of chemical plants dropped on
African American communities like a bomb."[28] Of course, activists argue that
Cancer Alley does not suddenly halt outside of the New Orleans metropolitan
area. There are plenty of industrial sites within Orleans Parish and neighboring
Jefferson Parish and even more plants immediately downriver from Orleans in
the parishes of St. Bernard and Plaquemines. The entire river corridor from Baton
Rouge to the Gulf is part of Cancer Alley in terms of exposure to industrial

emissions. The same river that brought slaves and monoculture to Louisiana has also brought its successor.

PLANTATION TO PLANT

Infrastructural pipelines and petrochemical plants are part of a historic continuum from plantation land use. The economies of scale of sugar plantations required large plots of land with proximity to the river. After the Civil War, the Freedman Bureau distributed small adjacent landholdings to freed slaves and extended family groups. They left intact the plantations, which in the second half the twentieth century were sold off. "The vast scale of industrialized agriculture necessary for the profitable cultivation of sugarcane laid the ground for the region's . . . transition to industrial petrochemical production."[29] Former "freetowns" now stand at the fence lines of industrial plants. "Following the Civil War their towns arose next to the old plantations, and the industry that followed later simply introduced another plantation culture of its own, low wages, minimal employment and the profits going as far away as Germany and Japan."[30] Names such as the Diamond, Trepagnier, and Good Hope Plantations became ideal sites in the twentieth century for bringing oil ashore for storage and refining. "This exchange of land use—'from plantation to plant'—has exposed local residents, many of whom descend from slaves, to the life-limiting and protracted threat of harmful pollution."[31]

Some of the industrial operators restored the ornate antebellum homes. Famous plantations such as Ashland–Belle Helene, Destrehan, San Francisco, and Aillet House were purchased by Shell Chemical, Amoco, Marathon Oil, and Dow Chemical.[32] Restoration efforts are aided by historic preservation grants as well as by the companies themselves. Such practices, note scholars, preserve the aesthetics of white supremacy. When I was on an April 2022 guided bike tour through Cancer Alley, the tour guide pointed out a plaque at the Destrehan Plantation outside of Norco thanking Amoco for restoring the big house. The river road hosting the sugar plantations was once called Millionaire Mile from the nineteenth-century Louisiana sugar industry. "All the beautiful clothes, balls, and furniture, they don't talk about the enslaved lives that made it possible," said Sheila Tahir, the tour guide.[33]

As with generalities, there are exceptions. In La Place, Louisiana, the nonprofit museum and childhood home of Kid Ory, an important Black trombonist in early jazz, was once a plantation home. The home was also the epicenter of the largest slave revolt in US history. In 2023, the museum was saved from closing by a grant from a company called Greenfield, which is trying to build a controversial grain elevator ten miles away in Wallace, Louisiana, that would equal the height of the Statue of Liberty. Greenfield's project is highly controversial. And the Whitney Plantation home in Wallace was restored by Formosa Chemicals and then sold to

FIGURE 10. Cutting Sugarcane by Hand in Louisiana. The arduous work led to a shorter life span than working on cotton plantations among enslaved laborers. In heavy sugarcane parishes, deaths outnumbered births, requiring more imported slaves. Photo by William Henry Jackson (1843–1942), courtesy Library of Congress, https://lccn.loc.gov/2016817573.

a New Orleans lawyer who turned it into the only museum told from the perspective of the enslaved workers and speaks of the particularly harsh conditions of harvesting sugarcane in South Louisiana.[34]

THE RISE OF LOUISIANA SUGAR

Enslaved workers were forced to clear the bottomland forest and build the levees. That was followed by grueling work in unshaded fields: first in indigo processing and then sugarcane, which was particularly backbreaking. In sugar parishes, enslaved workers outnumbered whites and experienced a negative demographic birthrate. In other words, population did not naturally expand there. It shrank. Life expectancy harvesting sugarcane was lower than even cotton.[35] Nutrition was also poor. Many workers were lost to disease or suffered injuries in the fields and in hot sugarhouses during fall grinding season.[36] The difficulty of producing sugarcane—a tropical monocrop imposed on a subtropical climate—demanded a frenzied, militarized pace, particularly during the fall harvest season. An early frost could destroy the crop.[37]

Prior to industrial plantations, sugar production was considered so laborious that sugar was reserved as an exotic luxury. Until the twentieth century, sugarcane laborers were engaged in the most dangerous agricultural and industrial work in the United States. Stalks were cut by hand and immediately ground to capture their juice before spoilage. The juice was granulated through intensive boiling and reduction that required skilled, enslaved "suciers." Geographically, particularly during the domestic period of the slave trade, Louisiana was used as a threatened destination by masters whose workers were not obedient. Treated as mechanical tools, enslaved people were discarded when they had exhausted their use. "For the enslaved, each day began with the ringing of a bell to herald roll call at the industrial sugar factory. The factory consisted of a boiling house, a steam-powered sugar mill, and chimneys spewing smoke and steam into the humid air."[38] In the sugar mill, alongside adults, children toiled under the constant threat of boiling kettles, open furnaces, and grinding rollers. Khalil Gibran Muhammad writes in "The 1619 Project," "All along the endless carrier are arranged slave children, whose business it is to place the cane upon it, when it is conveyed through the shed into the main building."[39] Housing for enslaved people was located along the central road amid fields of cane in close vicinity to the sugar mills. Numerous elements of everyday life, including the time it took for the enslaved to reach the mill from their quarters, were timed and recorded.[40] "Interviews with formerly enslaved people testify to the tendency for pregnant women to labor in the fields throughout their term. If an overseer wanted to whip a pregnant woman, they would notoriously dig a hole in the ground to protect the unborn asset. Some enslaved children were born in the very fields where they would grow up to work away their lives."[41]

Sugar granulation in Louisiana was perfected by a free man of color named Antoine Morin, whose kettle transformed Louisiana's economy. Morin was hired by Jean Etienne de Boré, who owned an indigo plantation at Henry Clay Avenue adjacent to present-day Audubon Park in New Orleans. Morin was a chemist and botanist who had emigrated from the French colony of St. Domingue (now Haiti) and built a reputation for sugar production and granulation at the Terreaux-Boeufs Plantation in St. Bernard Parish. He was a "distinguished alumnus of École de Paris." He laid out de Boré's plantation in the French Caribbean pattern and, "ladle in hand, supervised the crystallization of de Boré's first kettle of sugar. That inaugural 1795 sugar harvest demonstrated the possibilities of a successful industry in Louisiana. It required a cash outlay of $4,000 and 40 slaves (valued at $1,200 each), but it returned a profit of $5,000."[42] In 1796, the sugar crop yielded 100,000 pounds of sugar at $12,000 profit, or the equivalent of $3 million today.[43] Encouraged by de Boré's success, more Louisiana planters undertook sugarcane cultivation, and as early as 1797, more than 550,000 pounds of sugar were exported from New Orleans. By 1801, there were seventy-five sugar mills in Louisiana.[44] Sugar production stretched along the German Coast upriver from New Orleans.

Sugar plantations were also established along the banks of Bayou Lafourche, Bayou Teche, and Bayou Terrebonne and in the Red River valley of present-day Rapides Parish. Since waterborne transportation was the primary means of moving goods before railroads, nearly all the early sugar plantations fronted a navigable waterway. The sugar kettle used on the de Boré plantation now sits by the LSU chemical engineering building. The commemorative marker for the kettle reads, "Used by Jean Etienne de Boré in 1795 to granulate sugar from Louisiana cane for the first time, thus revolutionizing Louisiana's economy."[45] There is no mention of Morin or the brutal conditions of the plantation sugarcane industry.

Word spread of the quick profits to be made in sugarcane. After the United States acquired the Louisiana Territory in 1803, Anglo-Americans from as far as the Atlantic seaboard began arriving. Labor demands for sugar plantations also increased the Black enslaved rural population outside New Orleans. They outnumbered white residents by three to one, which took on its own specter of paranoia and oppression against revolt and uprising. By 1860, Louisiana planters were producing one-fourth of the world's sugar and were among the South's wealthiest slavers.[46]

HAITIAN INDEPENDENCE AND LOUISIANA SUGAR

But Louisiana's sugar high was also fueled by a massive slave revolt in present-day Haiti that dismantled the French colonial sugar monopoly. There is a direct link between Louisiana's ascendant sugar industry, Haitian independence, and the Louisiana Purchase. Some scholars argue that the Haitian Revolution led directly to the Louisiana Purchase, which by doubling the size of the United States set the young nation on course to become the dominant force in the hemisphere. From the second half of the eighteenth century until the French Revolution, the French colony of St. Domingue on the Caribbean island of Hispaniola was the richest colonial possession in the world. Many of the colony's early planters established hugely successful coffee, indigo, and sugar plantations. By the late 1750s, the enslaved population of St. Domingue numbered 500,000, which far outnumbered the white population of around 32,000. To establish social control, the French crown created a rigid caste system dominated by "grand blancs," who were French-born bureaucrats and landowners. White planters born in the colony were known as Creoles. Poor whites, or "petit blancs," formed an underclass. People of mixed ancestry and free men, known as *affranchise*, came next in the social hierarchy. At the bottom were enslaved Africans, who accounted for approximately one-third of the Atlantic slave trade.[47]

Despite the attempts of *grand blancs* to maintain control, violent conflicts between the enslaved and white landowners became more frequent, particularly as the revolutionary spirit mounted in France. The continental revolutionaries overthrew the French crown in 1789, and in March 1790, the newly formed

French National Assembly passed the Declaration of the Rights of Man and Citizen (Declaration des droits de l'Homme et du citoyen), which spread like wildfire throughout the French colonies. Eighteen months later, on August 22, 1791, the enslaved people of St. Domingue took control of an important northern province of the colony. When Revolutionary France declared war on England the same year, St. Domingue's white planters quickly agreed to support Great Britain. Most of the enslaved forces backed Spain, which controlled the rest of Hispaniola. Political rights through the idea of *liberté, égalité, fraternité* articulated by continental revolutionaries were legally extended to the free people of African descent in the French Caribbean by law on April 4, 1792. St. Domingue's white planters refused to recognize the decision. These tensions led to conflicts, initially between factions of whites and then between whites and free people of color. In January 1973, the French king, Louis XVI, was executed. Two months later, France declared war against the crowned heads of Britain, Holland, and Spain. On February 4, 1794, the French National Convention abolished slavery in all French colonies, "decreeing that all men, without distinction of color living in the colonies are French citizens enjoying all rights assured by the Constitution."[48] Slave revolts mushroomed throughout the Caribbean in 1795. Large slave owners and merchants abandoned France and the French Revolution.

Abolitionist sentiment also led to confrontations by working-class white Jacobin antiroyalists, who demanded equality for all people and abolition of slavery by any means necessary. French revolutionary Jacobins arrived on the shores of New Orleans inspired by the ideology of the *Rights of Man*. "They appeared in the smallest outposts, among the clergy, in all the city's taverns, and among the immigrant merchant community. French, St. Domingue, white, brown and black. Their precise contact with and influence among the slaves, though unknown, was a source of numerous White nightmares."[49] They also inspired paranoia and suspicion among Louisiana-Spanish customs authorities weary of the arrival of international agitators. In 1795, a plot was uncovered at the estate of Julien Poydras in Point Coupée Parish near New Roads, Louisiana, that involved enslaved and local whites along False River. Led by Poydras's overseer, Antoine Sarrasin, the rebels planned to set fire to several buildings and then use the confusion to seize weapons. They supposedly coordinated with individuals from neighboring estates to strike simultaneously. The plot also included a sympathetic white schoolteacher named Joseph Bouyavel, who had read the Declaration of the Rights of Man and Citizen aloud to enslaved men on nearby plantations. "Betrayed by informants among the Tunica Tribe, the plot was uncovered, and trials took place in May 1795."[50] The Tunica by the middle of the eighteenth century found themselves occupying a location between English and Spanish jurisdiction and attempted to curry favor with both.[51] The trial led to the convictions of fifty-seven enslaved people and three whites. "By June 2nd, 23 slaves were hung, their heads cut off and nailed to posts at several places along the Mississippi River from New Orleans to Pointe Coupée, where they were left for weeks. Others were severely flogged and assigned to hard labor in Spanish fortresses in Mexico, Florida, Cuba, and Puerto Rico. Three whites were deported." According to the trial documents, the conspiracy was inspired by the revolution in St. Domingue.[52] The conspiracy itself was part of a multiracial abolitionist movement supported by large segments of dispossessed populations in Louisiana and throughout the Caribbean. But conspiracy myths became the cornerstone to justify racist violence and oppression of anyone who opposed slavery and white supremacy. White schoolchildren were taught that the conspiracy proved that Black people were awaiting an opportunity to rise up and massacre all whites and take young white women as love slaves. Myths also targeted whites who opposed slavery as a danger to the survival of the white race. "It was used to enlist white Louisianians regardless of class to defend a racist system that was against the interest of the vast majority of the population."[53]

A major reshuffling of Western European powers had been set in motion. The unrest in St. Domingue led thousands of immigrants, including many free people of color and white planters, to move to Louisiana in the 1790s and early 1800s. Napoleon Bonaparte had hoped to retake St. Dominque and reestablish the island as a source of wealth for France. In 1800, he secretly negotiated with Spain in the Treaty of San Ildefonso to take possession of Louisiana, which would serve as a breadbasket for St. Domingue.[54] Separate attempts to retake St. Domingue in 1802 and 1803 failed. And by the end of the war, the French army was depleted. More than 80 percent of the soldiers who had been sent to the island had died. When French forces were finally defeated by Haitian rebels, Napoleon no longer saw a need for Louisiana. But he was also keenly interested in keeping it away from Great Britain. He sold the entire colony to the United States in April 1803. Thomas Jefferson's emissaries, Robert Livingston and James Monroe, negotiated with France initially for the sale of only the Isle of Orleans. Napoleon, who "realized he could use the sale to finance his campaign against Great Britain, offered to sell the entire colony for $15 million." Livingston and Monroe agreed to the purchase, and on December 20, 1803, Louisiana became US territory.[55] "Without the Haitian Revolution, it is unlikely that Napoleon would have sold a landmass that doubled the size of the then United States, especially as Jefferson had intended to approach the French simply looking to purchase New Orleans in order to have access to the heart of the Mississippi River."[56]

By then, in 1803, many of the largest plantations had already converted to sugar production to fill a huge void in the global market for sugar created by the Haitian Revolution. The large influx of St. Domingue immigrants helped further develop Louisiana's sugar industry. "Although sugar culture came late to Louisiana, it came fully formed, with planters and sugar experts fleeing the French colony of St. Domingue and the slave uprisings and civil wars that would culminate, in 1804, in Haitian Independence."[57] Antebellum Louisiana accounted for 95 percent of the sugar produced in the South. Louisiana's product was chiefly raw sugar, most of which was shipped to cities in the upper Mississippi Valley directly from the plantations or by way of New Orleans.[58]

The new US territorial governor, William C. C. Claiborne, was reluctant to allow St. Domingue refugees into Louisiana and wanted an end to the African slave trade. Marronage was also increasing with the booming sugar industry. "Instead of monitoring the entrance of the Mississippi River, Governor Claiborne would have done a better job of paying attention to the worsening working conditions in the cane fields and the sugar mills," writes Ibrahim Seck.[59] There was a growing fear of imported African slaves who had memories of prior lives, whereas slaves born in domestic servitude did not seem to instill the same threat. The United States banned the importation of slaves in 1808. Pleased with the ban, Claiborne still allowed some enslaved persons—referred to as "servants" on ship manifests—into the Louisiana Territory to appease planters' need for labor. Claiborne prohibited the immigration of free men of color but allowed passage of free women of color.[60] Louisiana Creoles with European lineages generally encouraged such immigration, seeing the refugees as potential cultural allies in the struggle against Americanization. Some of the Haitian immigrants became citizens of great standing in the community. New Orleans recorded the entrance of 9,059 French from Cuba in 1809 alone, which included 500 sugar planters, "whose 3000 slaves were authorized by Governor Claiborne to enter in spite of the U.S. prohibition against foreign slave imports into the territory."[61] Between May and July 1809, thirty-four vessels brought nearly 5,800 St. Domingue émigrés to Louisiana from Cuba. Immigrants from Guadalupe and other Caribbean islands soon followed.[62] The number of free people of color in New Orleans doubled, as did the number of French speakers in the city. But the grueling sugar plantation production and rising Jacobin abolitionist sentiment intersected to produce an atmosphere of smoldering resistance. For enslaved people throughout the rest of the New World, "the victory in Haiti—the story of which had spread through plantations across the south, at the edges of cotton fields and in the quiet corners of loud kitchens—served as an inspiration for what was possible."[63]

DESLONDES REVOLT

In all, ten thousand St. Domingue refugees arrived in Louisiana between 1809 and 1810. About one-third of them were white elites, another third were free people of color, and the remaining third were enslaved people who belonged to either the whites or free Blacks. The year 1810 had been marked by growing tensions of revolt in a series of regional upheavals. In Mexico, the Catholic priest Miguel Hidalgo sought to end colonial rule and abolish slavery. In the nearby West Florida parishes, which today are part of the state of Louisiana, a group of white landowners had just proclaimed an independent state from the Spanish regime. The fear of armed insurrection was in the air.[64] Louisiana officials were distracted by the explosive population growth that was taking place in just a few weeks. The disruption presented Charles Deslondes and his followers with an opportunity.[65] January 8, 1811 was cold and rainy. That night a group of enslaved men of the André plantation (now the Kid Ory Museum) surrounded the house, broke in, and killed André's son. Deslondes, according to court records, was André's overseer, an enslaved slave driver, who had been brought to Louisiana from St. Domingue. Armed with rifles, sabers, oak sticks, and other work tools, the rebels proceeded downriver toward New Orleans. They managed to set fire to property on adjacent plantations along the way and add recruits to their ranks. The plantation owner, André, was head of the local militia and escaped across the river to round up nearby planters.[66] By the next day, a stream of panicked whites entered New Orleans with breathless accounts of the uprising.[67]

Gen. Wade Hampton, a wealthy slaveholder from South Carolina, was quartering with troops in New Orleans to support the white-led revolt in Spanish West Florida parishes on the north shore of Lake Pontchartrain. Governor Claiborne ordered regular US Army troops under General Hampton to the scene of the insurrection. They joined with sailors from Commodore John Shaw's brig, the *Syre*, to total about a hundred men to confront the revolt.[68] On January 9, the governor activated the entire militia of the city and suburbs of New Orleans into service.[69] The number of rebels, meanwhile, had increased to several hundred men (eyewitness accounts vary widely from 150 to 500). Some wore uniforms. Drummers beat time as the rebel army marched downriver carrying flags and armed with knives, muskets, and machetes, chanting, "On to Orleans!," where they intended to establish a free territory. "Additional participants joined the ranks at the plantations of Labranch, Bernoudi, and Charbonnet, Butler, McCutcheon, Livaudais, and Arnould before marching on to the Destrehan plantation."[70] General Hampton referred to "roads half leg deep in mud," which the rebel army slogged through doggedly on their march downriver.[71] By that evening, one more planter was dead. Rebels made camp at Jacque Fortier's plantation, 25 miles outside of New Orleans. As Clint Smith points out, the relative discipline and organization of enslaved people, many of whom came from different countries in Africa and spoke different languages, is quite remarkable.[72] The conspirators had laid the groundwork for several months through secretive planning using coded language to avoid tipping off eavesdroppers.

The next day, having covered 6 miles, they arrived at the plantation of Jean François Trepagnier. Federal forces drew near at dawn on January 10. André's forces engaged Deslondes's group at François Bernard Bernoudi's plantation.[73] The US Army was well armed against the rebels, and the battle was predictably one-sided. In André's words, "We made considerable slaughter" (*un grand carnage*). About twenty-five enslaved rebels died in the attack. The rest dispersed into the cypress swamps, where they were tracked over the next few days. Deslondes's hands were chopped off, the bones in his legs were shattered by bullets, and he was burned over a bale of straw. A trial took place on the plantation of Jean

Noel Destrehan from January 13 to 15. A total of forty-five death sentences were administered without appeal, and twenty-two rebels received some kind of physical punishment. Each execution took place in front of the plantation at which that person was imprisoned. Heads were mounted on poles at the site where each was accused of committing his crime "in order to frighten all malefactors who would attempt any such rebellion in the future."[74] For months, the heads along the levee looked like "'crows sitting on long poles.'"[75] Mutilated bodies "would have flanked your path for miles—monuments to the savagery of the sugar interests and the irrepressible quest of enslaved people for liberty," writes Thompson.[76] In addition to those killed in action, twenty rebels were reported missing. By January 16, the planters requested an end to arrests.

Most of the slaves killed were between twenty and thirty years old and described as skilled suciers, which was a blow to the local labor pool.[77] In neighboring St. John the Baptist Parish, a separate trial led by Achille Trouard sentenced seven enslaved rebels to death. According to Seck, historian of the Whitney Plantation, much of this history is still being uncovered. St. Charles Parish has managed to preserve many documents (in contrast to St. John the Baptist and Orleans Parishes). The gravesite of François Trepagnier features a headstone in French saying that he was killed by "negro rebels," which testifies to the event. There are also primary documents in the Hill Library at LSU stating that planters would be reimbursed a third of the appraised value of each house burned during the insurrection. In addition, Governor Claiborne approved payments of $300 for each slave that was executed or killed during the rebellion.[78] Slaveholders were reimbursed for killing their slaves. General Hampton was "impressed" by the savagery of the Creole planter posse, writing "they are equal to the protection of their property." As a result, he was reassured to "invest heavily" in Louisiana sugar land. Louisiana's statehood was also on the line. "Claiborne hoped to assuage the fears of Louisianans and those in Washington, D.C., proving that he maintained control over his post and possessed the resources to suppress insurrection without external assistance."[79] Commodore Shaw likewise applauded the savagery as an example: "Had not the most prompt and energetic measures been thus taken, the whole coast would have exhibited a general sense of devastation; every description of poverty would have been consumed; and the country laid waste by rioters."[80]

CRACKDOWNS AND SLAVERY EXPANSION

Deslondes and his followers fell short of their goal. But they shook Louisianans to the core. "Despite the measures taken by Claiborne's administration and future leaders to tighten racial control, including increased patrols and a more vigilant militia, Louisianans never again felt complacent in their own safety." The 1811 revolt exposed additional regions of the country to the horrors of slavery. "Some northerners also used the insurrection to support their argument that Louisiana is not admitted into the United States, citing their fear that Louisiana posed an ideological and institutional threat to national stability and harmony."[81] The insurrection, and crackdown, helped spur the abolitionist movement. It is important to remember that heavy-handed enforcement of racial discrimination was in response to fear of losing territorial control. But there was never another slave insurrection in Louisiana. "To engage in open rebellion was tantamount to suicide," Smith writes.[82] He points out that our country's teachings about slavery, painfully limited, often focus singularly on heroic slave narratives at the expense of the millions of ordinary men and women: "Part of the insidiousness of white supremacy is that it illuminates the exceptional in order to implicitly blame those who cannot, in the most brutal circumstances, attain superhuman heights. It does this instead of blaming the system, the people who built it, the people who maintain it."[83] We forget the main lesson that the vast majority were regular people. This ordinariness is used to legitimate oppression, which is its own quiet violence.[84]

However, if rebellions were rare, white panic over alleged slave conspiracies were common. The records of colonial and nineteenth-century Louisiana are filled with documents alleging planned insurrections and "massacre of the whites," writes Fabor. In 1804, for example, New Orleans residents sent Governor Claiborne a petition alleging a plot by all the slaves in the city and insisting that he investigate and punish the guilty slaves "without any compassion."[85] Claiborne saw no evidence of a serious threat, and the panic passed—a trend that would be repeated. "According to the historian Brion Davis," Smith writes, "'for nearly 70 years, the image of Haiti hung over the South like a black cloud, a point of constant reference by pro-slavery leaders.'"[86]

Slavery, meanwhile, expanded dramatically in the Lower Mississippi Valley, driven by international demand for sugar and cotton and a growing domestic slave trade. After the importation of slaves was banned, a more nefarious rationalized and gendered brutality took hold. New Orleans became the domestic slave trading capital. By 1860, the number of enslaved people had multiplied sixteen-fold, with over 331,000 enslaved.[87] "By the time slavery was abolished in 1865, more than five hundred sugarcane plantations formed a seamless mosaic straddling both sides of the lower Mississippi River."[88] Until recently, most popular narratives lionized benign plantation masters and reminisced about the bygone era of the *lost cause* and luxurious living among the planter class. Despite the gap in the official historical record, oral histories of the revolts were shared by Black families of the area.[89] Margie Richard, a community activist who led a successful campaign for a fair buyout of homes contaminated by Shell Oil, famously tapped into the Deslondes revolt with the rejoinder recited by local organizers: "If my ancestors were willing to be killed for standing up to slavery, I can surely stand up to Shell."[90]

FREETOWNS AND ENVIRONMENTAL REDLINING

After the Civil War, the brief period of Reconstruction spawned "freetowns" of former plantation workers who were ceded parcels by the US Freedman's Bureau from the very plantations they worked. They may have been in the same cottages along the side or central thoroughfare in the rear. W. E. B. Du Bois wrote in 1901 that the Freedman's Bureau was the government's attempt—however fleeting—to answer the deeper issue of race in America that the Emancipation Proclamation seemed to intensify: "What shall be done with the slaves?" The Freedman's Bureau was created by Congress in 1865 and closed in 1872.[91] It was directed by Oliver Otis Howard, a Union general whose name adorns Howard University; he sought to transition four million newly freed slaves into a free-labor society. "An honest and sincere man, with rather too much faith in human nature, little aptitude for systematic business and intricate detail, he was nevertheless conservative, hard-working, and, above all, acquainted at first-hand with much of the work before him," wrote Du Bois.[92] Many of these freetowns became historic African American hamlets; and later easy marks for industrial expansion.

In St. James Parish, a hamlet by the name of Freetown was founded by former slaves in 1872.[93] Situated on the former Pedescleaux-Landry sugar plantation, it was once a bucolic agricultural community. Its degradation happened relatively slowly, as petrochemical plants began moving in next door. Eventually, residents became stranded in place. Their homes were stripped of the characteristics that made them livable. This, according to Rob Nixon, constitutes the slow violence. There are many such examples of multigenerational residents finding themselves today squeezed between industrial behemoths on former fields and forests that were transformed by heavy industry before their very eyes.[94] "Ecological degradation, climate change and cancer risk are contemporary inheritance, and by-products of colonial genocide and slavery," writes Forensic Architecture, a UK-based firm that advocates for environmental justice communities.[95] Currently, there are proposals to build or expand three plants in St. James Parish, where 14 percent of the land is owned by chemical companies and over 40 percent of the parish is wetlands.

In 2014, the St. James Parish Council passed a comprehensive land-use ordinance that quietly rezoned large portions of its predominantly Black 5th District from residential to residential/industrial, which is "a particularly pernicious type of zoning ordinance," where industrial facilities and residential homes stand side by side without adequate buffer zones.[96] The introduction of the land-use ordinance initially included language reminiscing about a bygone era of "luxurious living and delightful ease" when acres of land "were counted by thousands and slaves by hundreds."[97] Several St. James Parish Council members, including the president, are current or former employees of the petrochemical industry. The plan rezoned the majority-Black community of Burton Lane as "industrial" and the majority-Black community of Welcome as "existing residential/future industrial." Data collected in the 5th District shows that residents have been exposed to emissions that can reach 765 times the levels considered safe by the EPA.[98] In the single community of Convent, a small town on the Mississippi River in St. James Parish between Baton Rouge and New Orleans, eleven chemical plants lay within a few miles of the town, which is 80 percent African American with a 40 percent poverty rate.

In April 2022, the Environmental Protection Agency under President Biden said it was investigating whether two state agencies discriminated against Black residents when they approved permits for two chemical plants and a grain terminal in St. John the Baptist and St. James Parishes. The probes followed a visit in January 2022 by the Biden-appointed EPA administrator Michael Regan, who promised a crackdown on racially biased permitting decisions in Cancer Alley. The probes focused on the approvals by the Department of Environmental Quality and the Department of Health and Human Services for three major projects that would produce likely carcinogens of chloroprene and ethylene oxide,[99] as well as volatile carbon monoxide, benzene, formaldehyde, nitrogen dioxide, and particulate matter called PM2.5.

According to the EPA, residents at the census tract near one of the plants owned by Denka experience the highest risk of cancer in the United States.[100] In response to the complaints, Denka shot back that there were no widespread elevated cancer rates in St. John Parish compared to the state average, based on the Louisiana Tumor Registry. "The complaint (filed against Denka) claims local, state and federal officials have turned a blind eye to health impacts in the area, but in fact these agencies have been studying the situation long before these groups got involved—and choose to consider real science rather than sensational pseudo-studies."[101]

LULUS

Proposals to site a new industrial facility are often sold to a community in terms of the jobs it will create. This tradeoff may be viewed as a kind of economic blackmail foisted on hardscrabble neighborhoods with otherwise few job opportunities.[102] The residents of Cancer Alley are repeatedly told that fighting the influx of new plants will kill jobs, as if they must accept the industry that poisons them. "Residents along fence lines with heavy industry often experience elevated rates of respiratory disease, cancer, reproductive disorders, birth defects, learning disabilities, psychiatric disorders, eye problems, headaches, nose bleeds, skin rashes, and early death."[103] Air pollution is compounded with "the stress and tension of noise and squalor."[104] What was sold as an economic booster becomes an economic drag. Residents become entrapped by poor real estate equity in undesirable areas, which land-use professionals call "Locally Unwanted Land Uses," or LULUs. Study after study affirms that minority communities face an uneven pollution burden compared to white counterparts, even those with lower incomes. For example:

- Researchers from the University of Colorado at Boulder reported in 2008 that African Americans with household incomes between $50,000 and $60,000 live in neighborhoods that are, on average, more polluted than the white households with $10,000 less earnings.[105]
- African Americans are much more likely to live near toxic pollution and are exposed to 38 percent more air pollution than white Americans.[106]
- Likewise, in neighborhoods with "clustered facilities," people of color make up 69 percent of nearby residents. Such disparities were repeated in nine of ten US EPA regions and 40 of 44 states with hazardous waste sites.[107]
- Fines levied on polluting industries are also unequal. An examination of 1,100 Superfund sites reveals that the average fine imposed on polluters in white areas was 506 percent higher than the average fine imposed in minority communities.[108]
- Meanwhile, white residents experience a "pollution advantage" in the goods they consume, according to a 2019 study in the *Proceedings of the National Academies of Sciences*. Whites consume more goods and services than Blacks and Hispanics, yet are exposed to less air pollution associated with the production of those goods and services. Meanwhile, Blacks and Hispanics bear a "pollution burden" by consuming more air pollution and fewer goods and services. "Low-income and minority populations, living adjacent to heavy industry and military bases are required to make disproportionate health and economic sacrifices that more affluent people can avoid."[109]
- In St. James Parish, a GIS study found that polluting industries were located in areas with the highest percentages of African Americans, the lowest average household income, and the most residents without a high school diploma. Meanwhile, the residents employed by the plants tended to live the farthest away, were wealthier, were better educated, and were more likely to be white.[110]

A recent peer-reviewed academic study from Tulane University found direct correlations between cancer incidences and cancer risk in Cancer Alley census tracts. Researchers controlled for parish-level smoking and obesity rates. They pulled the most recent cancer data from the Louisiana Tumor Registry (2008–17) and estimates of race, poverty, and occupation from the US Census Bureau's American Community Survey (2011–15) and estimated cancer risk due to emission point sources from the EPA's National Air Toxics Assessment.

They found a statewide link between cancer rates and carcinogenic air pollution in marginalized communities. "These findings are consistent with the firsthand knowledge of Louisiana residents from predominantly Black, impoverished, and industrialized neighborhoods who have long maintained that their communities are overburdened with cancer."[111] This reaffirms what activists of color have been saying for years: environmental racism is not one of slurs and epithets; it is more subtle and more powerful. It's racism of neglect.

While they bear the brunt of pollution, environmental justice communities likewise suffer from inadequate protections from the Louisiana Department of Environmental Quality (DEQ), which is charged with regulating one of the busiest clusters of industrial activity in the world, is perennially underfunded, and is reliant on archaic methodology and self-reporting by the polluters themselves.[112] In January 2021, the Louisiana Legislative Auditor issued a scathing report of the inadequate enforcement practices by DEQ, which took nearly twenty months to issue enforcement actions after a plant operator failed to properly report emission violations. "Auditors also found it could take as long as nine years from the time a company was cited for violating emission standards before it was ordered to pay a fine or had a settlement approved requiring the company to pay for a mitigation project."[113] That means the plant could excessively pollute for more than a decade before being forced to stop. The department also doesn't adequately track the penalties it has assessed or whether penalties were even paid. Plant reports are mailed to DEQ and then manually scanned into the agency's database, which the audit says results in unreliable reporting on when and whether the reports were received.[114]

The department also only considers the effects of individual chemicals emitted by plants, instead of considering the holistic impact of their combined effects. Nor does it conduct or mandate regular air monitoring. Meanwhile, there is also little granular air monitoring by federal or state agencies. Instead, the EPA models a "Risk-Screening Environmental Indicator" database to identify potential high-pollution areas. Plants with emissions that exceed certain threshold levels are required to report them to the government. "According to EPA data, the number of industrial plants in Louisiana that reported their toxic releases grew from 255 to 320 in the last three decades, an increase of 25 percent."[115] Louisiana has the highest toxic air emissions per square mile of any state, based on data gathered by the US Environmental Protection Agency's 2018 Toxics Release Inventory. The state in 2018 averaged 1,239 pounds of toxic air released per square mile, well ahead of runner-up Ohio, with 899 pounds per square mile.[116]

Louisiana also far out-spills every state in the nation, magnifying exposure of co-pollutants to marginalized communities.[117] And at least thirty facilities in Louisiana marshes contain the most toxic chemicals allowable by the EPA, which makes them particularly dangerous in hurricanes. During emergency shutdowns that happen with regularity during the fall hurricane season, for instance, harmful "spot plant flaring" legally releases tons of pollutants that magnify exposure of the toxic payloads inside plants. The plants themselves are structurally at risk from storm damage, often unbeknownst to nearby residents.[118]

HURRICANE IDA

Even as I write this, fence line communities are still recovering from the most powerful hurricane to ever hit Louisiana. Category 5 Hurricane Ida ripped through the Mississippi River corridor on August 29, 2022, sixteen years to the day after Katrina. Ida raked across hundreds of chemical facilities and left a million people without power. Half of the nation's petroleum refining and natural gas processing capacity is based along the Gulf Coast.[119] Ida left behind miles of visible oil slicks in coastal waters.[120] Power outages knocked offline seventeen state air monitoring sites, as well as the hotline for the Louisiana State Police, whose hazardous materials unit handles toxic emissions. Still weeks after the storm, blazing flair-offs and black smoke could be seen belching from smokestacks, which is associated with emergency releases. The lack of electricity at other refineries and the inability to supply steam and nitrogen to the massive flares indicated that toxic chemicals were being improperly released without burn-off.[121]

After Hurricane Laura made landfall near Lake Charles in August 2020, it took several days for the DEQ to deploy mobile air monitors.[122] This is the context of vulnerability that residents face. Adding insult to injury, Louisiana's perverse Industrial Tax Exemption Program removes 80 percent of a plant's value from local tax rolls for up to ten years, depriving communities of critical tax resources for libraries, schools, and clinics.[123] Residents in Cancer Alley live in a perennial state of emergency from catastrophic storms, floods, toxic releases of chemicals, and spills. Such trauma is compounded by stay-indoors orders, negligent oversight, and poor health outcomes and morbidities from diseases like COVID-19. People with the least resources and legal knowledge to resist are burdened with presenting a cogent case of harm against powerful actors with the backing of the state regulatory apparatus. This is a particular challenge when environmental degradation appears naturalized in the form of floods and storms, or as invisible toxins in the air and water. Such embedded, imbricated effects can be easily veiled. In effect, any earnest attempt to correlate hazards with emissions and their impact on communities is left to the enterprising solutions of residents themselves, who document excessive chemical releases using low-tech EPA "grab bucket" kits, whose samples must be tested in a private lab. The samples add legitimacy to residents' stories of respiratory illnesses, skin irritations, cancer, stillbirths, and general trauma. With the tools of citizen science, they have managed to challenge assertions by powerful authorities

FIGURE 11. Rise St. James founder, Sharon Lavigne, protests the state's approval of a $12 billion plastics facility proposed by the Taiwan-based Formosa, which purchased 2,400 acres of the former Buena Vista Plantation. Her efforts not only halted the project, but she was recognized with the 2021 Goldman Environmental Prize for her grassroots organization work. Photo courtesy of Steve Pavey.

who downplay emergency events. "Climate change to them is not an existential threat," says Tahir, who leads social justice bike tours for the Louisiana Bucket Brigade. "They're in the trenches."[124]

RESISTANCE IS MOUNTING

A renewed sense of energy has taken hold among residents who are publicly and forcefully protesting plans by companies with powerful backers to build or expand operations. One of the region's community organizers, a retired schoolteacher, Sharon Lavigne, said she just got tired of the lies and growing sickness. She began organizing against the $12 billion Formosa plastics project in St. James Parish. That was in 2018. The efforts of her group, Rise St. James, has so far managed to notch several victories. "I felt desperation living in a world full of pollution and people dying," Lavigne explained. "Homes depreciated. Our way of living changed since industry invaded us."[125] Lavigne, who was recognized in 2021 with the Goldman Environmental Prize, the second recipient in Cancer Alley, said the governor and local parish and port officials welcomed Formosa without notifying residents. The Taiwan-based company purchased 2,400 acres on the former Buena Vista Plantation.

According to its air permit application, Formosa would be the single largest emitter of ethylene oxide and benzene (both carcinogens) in a state that averages the highest toxic air emissions per mile in the country. "They aren't informing the community about toxic air pollutants that will poison us. . . . We have to defend our community."[126] Lavigne's organization partnered with the Louisiana Bucket Brigade, Healthy Gulf, and the Center for Biological Diversity to successfully petition the federal government to halt permit approvals for the Formosa project, which was dubiously named the "Sunshine" project. The Formosa footprint along the west bank of the Mississippi includes chemical complexes, ship and barge docks, railroads, and power lines adjacent to the river, as well as wetlands. The lawsuit by Rise St. James cites several environmental factors: loss of wetlands in construction, pollution of single-use plastics, and toxic emissions. It also alleges the Army Corps of Engineers failed to consider environmental damages and public health risks under the National Environmental Policy Act. Petrochemical companies are also required by federal law to identify historic properties and cemeteries threatened by development. Rise St. James alleges that the Corps failed to adequately consider the harm to slave burial grounds on the site, which is a violation of the National Historic Preservation Act: "The Plastics Facility is sited on two 19th century sugarcane plantations, which include two cemeteries that contain the remains of enslaved people."[127]

Plants typically hire for-profit archaeological firms, who are incentivized to produce reports favorable to plant owners.[128] "If they find positive evidence of remains there, that puts a sinker torpedo in their plans," said Justin Kray, a cartographer.[129] Formosa's private archaeology firm, Cox-McClain, dug near but missed four gravesites of the Arcadia cemetery, which is known to the community. The Formosa archaeologist relied solely on a single historical survey that omitted most of the cemetery remains of the period. The 1894 Mississippi River Commission report is a poor index for what the firm was supposedly investigating, according to the researcher Imani Jacqueline Brown. An investigation of the US Coast Survey Maps of 1877 and 1878 lists several markings of cemeteries on the Formosa footprint, located at the Eline, Acadia, and Buena Vista Plantations. "The [Cox-McClain] firm came to the conclusion that there were not cultural resources on site that should impede Formosa's development."[130] A simple cross-reference of the other two maps would lead to a substantively different conclusion. However, such due diligence rarely occurs. Brown, who is a researcher with Forensic Architecture, studied over fifty field reports from different for-profit archaeologists and found a systemic lack of regard for antebellum Black cemeteries. In 2015, a cemetery with a thousand artifacts was uncovered on the Monroe/Houmas and Bruslie Plantations, now owned by Shell Oil Company, during a survey for a proposed expansion of the Shell Convent Refinery. For decades, Black residents of neighboring communities had attempted to alert archaeologists and state officials to the locations of these cemeteries.[131]

RECLAIMING THE NARRATIVE

On September 14, 2022, a state judge tossed out Formosa's air permits, telling state regulators to start from the beginning. In a thirty-four-page ruling, the Nineteenth Judicial District judge said the department wrongly approved the permit without conducting a full environmental justice analysis to see if the plant disproportionately harmed minority communities.[132] Meanwhile, the Corps of Engineers announced on November 14, 2020, that it would reevaluate its wetlands permit for Formosa Plastics. In the intervening period, Rise St. James's archaeology consultants confirmed evidence of a slave cemetery on the property. On August 18, 2021, the acting commanding general of the US Army Corps of Engineers ordered the Corps to prepare an Environmental Impact Statement to assess Formosa's potential impacts on the quality of the human environment in the region, potentially delaying the project for years. "The Army Corps has finally heard our pleas and understands our pain. With God's help, Formosa Plastics will soon pull out of our community," said Lavigne. "Nobody took it upon themselves to speak for St. James Parish until we started working to stop Formosa Plastics. Now the world is watching this important victory for environmental justice."[133]

Brown and Forensics Architecture say there is likely a cemetery for enslaved people at each of the five hundred plantations that lined the river before the twentieth century. Comparing satellite imagery with earlier aerial photographs and historical surveys, some of which were hand-drawn, researchers can reconstruct a landscape that once existed. They focus on anomalies in the landscape that may represent ruins of sugar mills or groves of trees that marked where enslaved people were interred. A basic understanding of "plantation spatial logics" would point researchers to likely sites. Colonial land grants issued territory by parcels marked in 40 to 80 French arpents, or 1.5 to 3 miles, from the river to the back swamp. The land would have been accessed by a main plantation road. The industrial sugar mill and the enslaved quarters would reside on this grid to the rear. Larger plantations may contain two cemeteries, one at the end of the first arpent and one at the end of the edge of the second arpent. Enslaved cemeteries were near the back swamps. As we have read, those swamps were drained during periods of increased cultivation, so some remains became marooned in the middle of cultivated fields, which can now be identified as topographic anomalies. Over time, a sacred grove becomes part of the landscape. "Logic tells us that if people were enslaved on this land and they died on this land, chances are they are buried on this land," said Lenora Gobert, a genealogist with the Louisiana Bucket Brigade. "Every plantation has a cemetery. They lived there, they died there, they are buried there."[134] While state archaeologists often identify cultural resources on high ground near river levees, enslaved interments are more likely to occupy the lower "liminal zone" behind plantation houses near or within the forests. "While we have no data on cemeteries beyond the forests' edge, it's probable that cemeteries exist within them," says Brown. Anomalies within these "zones of probability" should be investigated for possible remains.

While state law allows a company to dig up the historical remains and move them, the local activists believe that enough negative publicity will push policy makers and local council members to reevaluate their support of such projects. They also hope to apply pressure to the financial backers that issue bond sales for them. "Our work is to craft the narrative," said Anne Rolfes, founder of the Louisiana Bucket Brigade. "Do you think Formosa is going to try to dig people up? That would be a serious error."[135]

Much history of African American freetowns is yet to be written. It has behooved the plant owners to obfuscate historical claims these communities have to the area. With some exceptions, local tourism projects reconstruct plantation homes and a benevolent spin on the brutality of a slave economy. Yet new artifacts are being documented despite resistance by corporate landowners. Such knowledge here is power. Formosa, for example, had been told to fence off the gravesite discovered in 2015 and maintain a certain distance from it but was captured on drone footage violating the stay-away order. Subsequently, the Louisiana Legislature passed a law in the 2021 session banning drones from pipeline operations. "It's a response to what's happening. Every punch we give, they run to the legislature to implement dictatorial tactics that won't work," said Rolfes. "It shows our strength."[136] This history faces hostile forces. Forensic Architecture identified 1,200 possible interment sites from a mosaic of 1940 aerial images. Fewer than 350 of them remained in 2021. The Louisiana Department of Economic Development identifies 200 "development-ready" sites along the river. Those in areas slated for development need to be urgently investigated, writes Brown. "Moreover, the entire landscape is revealed as holding historical and cultural value that we can still recover."[137]

This ongoing struggle highlights historical connections to the political economy of plantation agriculture and the contemporary petrochemical industry that replaced it. It also serves as a hopeful coda to community efforts to reclaim the notion of "public good" through civil disobedience, marches, legal claims, and public pressure on banks that fund the fossil fuel industries. Armed with buckets and bullhorns, residents of Cancer Alley have transformed their private suffering into a public battlefront. On the streets, in the courthouses, in the media, their confrontations are breaking the historical cycle of racialized violence. Louisiana already ranks number 1 in the nation of per capita petroleum consumption because of its petrochemical and oil and gas industries.[138] Residents are writing historical counternarratives to the pastoral antebellum South to help return sovereignty to those who have lived and labored here for generations. "If we keep that logic in mind on this process it makes sense of what we are trying to do," Gobert said during a Zoom workshop. "W. E. B. Du Bois once said the Negro has no history. In 2021, if the Negro has no history, he's expendable. We're showing he is not expendable. We are building a cultural defense against the petrochemical industry."[139]

Interlude

Landfall

I read on Twitter that waiting for a hurricane is like being stalked by a turtle. There are intense preparations, followed by extended lulls and worried second-guessing. It's an autumn ritual here: watching for tropical disturbances to "organize" into named storms, which increases potential for danger and activates higher insurance deductibles. The turtle crawls to the Gulf, having ravaged a Caribbean outpost before reassembling and strengthening again. Its Doppler-surveilled path is dotted by text alerts. Local stations interrupt their regularly scheduled programming. Meteorologists perfect the art of repetition. Reporters stake out their usual spots: at Lake Pontchartrain to measure waves; on the Gulf Coast to interview the fool-hearted and fearful; at crowded grocery stores; at sandbag distribution points; over highway evacuation routes. From podiums, officials urge precautions. Schools announce closures. We stare at the Doppler's latest rotation. Family members elsewhere check in with offers of spare bedrooms. We haul plants inside, tuck away patio furniture, and tie down the porch swings. Check the corner bodega for freezer ice and batteries. Find the hurricane box of candles and nonperishables in the closet. Top off the gas tank and park on higher ground. Canvas the neighborhood. Clear debris from storm drains. We will discuss taping up the windows but won't do it. Pull down the shades. Charge cell phones. Locate board games and playing cards. Watch the turtle.

As it makes landfall, we see split screens of Doppler radar and wind- and rain-battered communities: swinging traffic lights, swaying trees, and torn away rooftops. Outer bands of weather test our defenses. The sky darkens. The wind whips through in waves like timed contractions. Eventually, cells of extreme gusts begin lashing against the house. When the power goes out, which it will, we track local broadcasts for a bit on cell phones and then a transistor radio. We keep playing board games. At night, the electric lamps go on. The wind rips against the house. We move away from the windows. We present calm faces to the children. And when a particularly long gust lashes against the window and you hear something crack or break, you feel your body clench and your abdomen clamp and, momentarily, you feel alone.

Conclusion

Stuck in the Mud

William Burrows famously wrote in his 1959 countercultural novel, *Naked Lunch*, about the stark reality that slides underneath consumptive capitalism: it describes "a frozen moment when everyone sees what is at the end of every fork." We can pause to bring to light the contradictions of American capitalism and still not succumb to despair. Grief can turn to righteous anger. We can identify how reputations are greenwashed. We may applaud the care by a "survival center" sponsored by Freeport McMoran for animals injured by oil spills, but also recognize that Freeport McMoran is an international oil and mining company that in 2021 agreed to settlement talks in a $100 million environmental lawsuit for its role in pipeline-related destruction in the Louisiana marsh. We must keep these two opposing ideas in our heads.

It's part of living here at the edge of doom. Yet doom cannot equal hopelessness. The game is not up. In fact, it has just begun. When we find the future too dystopian and the needed changes too draconian, it may help to follow the provocation by Haraway to "stay with the trouble." In other words, we stay with one another and work *with* rather than *on*.[1] We can and should push back against the narrative of hopelessness. Science communicators themselves argue that the report by the United Nations Interagency Panel on Climate Change (IPCC) offers some immediate opportunities. Even as ominous warnings about methane gas releases from permafrosts point toward the dreaded positive feedback loop, buried in chapter 6, the 2021 report actually shows that immediate reductions in methane emissions and other short-lived climate pollutants could reduce warming 0.2 degree Celsius by 2040 and .08 degree Celsius by the end of the century. That's pretty substantial in the context of 1.5 degrees Celsius outlined by the 2015 Paris Agreement. The IPCC report also stated that methane and other short-lived pollutants released in 2022 will have "at least as large" an impact on climate change over the next ten to twenty years as current carbon dioxide emissions. Their extreme potency, combined with their limited time in the atmosphere, means that reducing those emissions will have a significant and "almost instantaneous" climate impact.[2] Other collaborative scientific reports support this assessment.[3] Framing the situation as hopeless saps the will to make necessary changes today for human survival. It also saps our capacity for imagining and caring for others' worlds. It defeats our ability to make meaning and becoming with each other and other "companion species" who will continue living and dying. "To think-with is to stay with the natural-cultural multispecies trouble on Earth," Haraway reminds us. Our present and unfolding future in the ruins is "relentlessly contingent" on what we do and on what kinds of "storying and worlding" we can do with familiar and unorthodox agents.[4]

We may look to Anna Tsing's investigation of Matsutake mushroom picking, which reveals something powerful about flourishing among ecologically harmed areas. The Matsutake circulates globally through an exchange system that defies traditional economics.[5] By following the Matsutake, Tsing discovers assemblages of Japanese buyers and American businessmen, as well as pickers from multiple communities: Chinese, Korean, Hmong, Lao, and Mexican, as well as disaffected Americans. Getting mushrooms to mouths requires informal exchanges along a network of pickers, buyers, shippers, restaurateurs, and businessmen. Tsing also attends to the scientists, foresters, DNA sequencers, fungal spores, trees, and mycorrhiza. She refuses to reduce the Earth's urgency to abstract destruction. "She looks for the eruptions of unexpected liveliness and the contaminated and nondeterministic, unfinished, ongoing practices of living in the ruins." In other words, by following the Matsutake, Tsing is guided to possibilities of coexistence within environmental disturbance. "Matsutake tell us about surviving collaboratively in disturbance and contamination."[6]

What about possibilities for monitoring air pollution using racing homing pigeons, who can traverse heights inaccessible to more expensive ground-based monitors? "These data could also be streamed in real time to the public via the Internet. What would it take to enlist the cooperation of such birds and their people, and what kind of caring and response-ability could such a collaboration evoke?"[7] Pigeons in this case are not inert transistors or SIM cards but coproducers that interact with and impress on their co-trainers. There's a kind of cohabitancy in the ruins—characterized in how different beings come together in ways that can only happen through the "unnatural" disturbances of one another's native habitat. Haraway calls these multispecies interactions "flawed translations across difference" that allow for a modest flourishing.

In these cases, emerging natures are coproduced through disturbance. "Out of the ruins of the Anthropocene might then emerge 'wildness' as an inclusive principle that meshes the organic being that inhabits us with the environment that surrounds us."[8] We cannot afford to denounce a kind of world in the name of modern nostalgia. The sanctification of a natural world separated from humans simply reinforces the fallacy of a pristine untouched wilderness and, ironically, reinforces poor stewardship and ignorance that we are part of its ecology. Life continues to assert itself among the ruins as wildness, instead of wilderness. Wilderness is romantic and limited. Wildness is rather quite ubiquitous. While pristine wilderness barely exists, wildness persists in all kinds of uncharted spaces.[9]

At the river, wildness is displayed by purple irises and pink dandelions that sprout through the concrete revetment levee plates; turkey vultures perched atop trees that have been thinned by repeated hurricanes. Wildness is the breaking through; trilling insects in the city; frogs just out of sight in the evening; crickets rubbing; beetles in the bedroom; a puppy munching on a cockroach on his walk. My repulsion notwithstanding, wildness resists or rather overwhelms our controlling impulses. It flourishes despite them: when fields are left to "fallow" or offices are left to the woods. There is an ironic bit of comfort in that. It arcs beyond the Anthropocene to an unfolding that collects and layers its trash and chips into a planetary compost like a "mad gardener."[10]

GREENWASHING NET ZERO

But we have work to do. We must be vigilant and skeptical when the fossil fuel industry manages to embed itself in the solutions for achieving a sustainable future. Not only in the Louisiana Master Plan, but in any narrative of "having a seat at the table." Lately, the industry line has targeted achieving net zero through industrial carbon capture and storage (CCS), which is embraced by elected leadership—despite the failure of carbon capture pilot projects. Over 80 percent of the thirty-nine projects that have sought to commercialize carbon capture and storage have failed.[11] "Industry wants to build a whole new network of pipelines through our wetlands, through Cancer Alley and other environmental justice communities, and they want billions and billions of dollars not only in federal money but state subsidies too. Because there is no money to be made in the marketplace," says Rob Verchick, a member of Governor Edwards's Climate Action Task Force and president of the Center for Progressive Policy Reform. "The only money to be made is if you can get the federal government to pay you."[12]

Biden's bipartisan infrastructure bill and Inflation Reduction Act, for example, increased the "45Q" tax credits for carbon dioxide that is either stored underground or used to bolster oil extraction.[13] According to Verchick, the federal government is subsidizing $12 billion in carbon capture spending with only a quarter going to ambient air capture. "The problem is that despite billions spent on it, the CCS industry has failed to even meet its own expectations." The Government Accountability Office reported in 2021 that the federal Department of Energy invested a billion dollars in nine different CCS projects between 2010–2017 and only two were still operating. "It's an amazingly expensive way for oil and gas companies to keep burning fossil fuels," says Verchick.[14] Over five hundred environmental organizations signed an open letter, published in the *Washington Post* in July 2021, calling carbon capture and storage a false solution. "We already have a huge problem with methane and leakage."[15]

Louisiana has historically positioned itself as a fossil fuel–friendly state while it has increasingly become a case study for environmental racism. These two elements are part of the same legacy. Carbon capture advocates seem to be grasping the first part of the equation without its negative externalities. More industrial development would increase the potential for plant accidents, water contamination, noise, and squalor that disproportionately affect environmental justice communities. It is likely to rationalize demand for new projects running on dirty energy sources, delaying the energy transition to renewables, extending the life cycle of aging plants, and misdirecting critical public resources that could be spent more productively elsewhere.

Despite the vast public relations and marketing budgets promoting carbon capture, the road to net zero based on modeling by the governor's own Climate Action Plan requires electrification of the industrial grid, expansion of forests, and no-carbon hydrogen.[16] Carbon capture was modeled at best case to nominally reduce greenhouse gas emissions if it was tied to production activity with no carbon sources and other combinations of natural carbon sequestration, energy transition strategies, and conservation. "It's less than 10 million metric tons. It's a rounding error almost. If you compare that to the potential harms and all the money, it doesn't seem like a very good investment," explains Verchick in a zoom presentation, "The False Promise of Carbon Capture."[17] This rounding error also assumes competent oversight by a regulatory apparatus with a poor track record of enforcement. In a technical memorandum accompanying the state's Climate Action Plan, the Water Institute of the Gulf, which facilitated task force collaborations, argued that industrial carbon capture alone would have minimal effect on reducing greenhouse gas emissions. Louisiana is a carbon-intensive producer. Most of its energy consumption—over 70 percent—comes from industrial refining, chemical manufacturing, and natural gas processing.[18] The state ranks number two in per capita energy consumption, number one in per capita petroleum consumption, and forty-ninth in renewable energy consumption, according to the Energy Information Administration. Unless Louisiana's intensive industrial sector transitions to fully electric, clean energy through fuel switches and heightened efficiency, it will do little to lower emissions.[19]

Though it sounds green, industrial carbon capture does not remove legacy carbons from the atmosphere as forests and oceans do. It works by capping new emissions at the smokestack and transporting them by high-pressure pipelines to other sites where they can be contained in deep geologic formations, supposedly forever. It requires new infrastructure and pipelines, which would slice up more of Louisiana's degraded environment. It also, presumably, continues to expose environmental justice communities that currently bear the burden of industrial production, since many of the same facilities would continue operations. It would encourage new construction and buildout along sites considered shovel ready in the crowded Cancer Alley industrial corridor. It would require billions of dollars of public subsidies.

But the promise of local capital investments, construction jobs, and economic development has officials applauding a slew of billion-dollar (tax-exempt) projects that sound too green to be true. "We're a natural fit for it. This is where capital investment is going to continue to flow," said Governor Edwards.[20] Indeed, industrial carbon capture and storage seems to crop up whenever it comes to discussing capital investments and job creation. Louisiana's Climate Action Plan cites a study by the Rhodium Group that touted the potential job boom for carbon capture projects. It said carbon capture projects could produce between 2,700 and 4,000 total jobs per year on average over the next fifteen years and an additional 1,700 to 2,500 jobs per year to operate and completely retrofit older plants. This suggests carbon capture is more about job creation—even if subsidized—than reducing carbon footprints.

Officials and lobbyists touting carbon capture projects are not sharing much information with fence line communities, who are weary of aging, accident-prone plants. CO_2 is a highly corrosive material, says Monique Harden with the Deep South Center for Environmental Justice. In 2020, a pipeline carrying compressed carbon dioxide ruptured in the town of Satartia, Mississippi, which caused over forty people to get hospital treatment and more than three hundred to evacuate. The incident illustrates the potential dangers of transporting carbon over long distances. "Right now, we don't have a legal way of mitigating environmental racism. You don't see community folks talking about carbon capture sequestration or carbon capture storage," said Harden.[21]

A company called Gulf Coast Sequestration, which had a representative on the governor's Climate Task Force, has applied to the US Environmental Protection Agency to inject carbon into a deep reservoir in southwestern Louisiana that is already under stabilization protocols for environmental damages. The Louisiana Department of Natural Resources filed a permit to regulate that site—called Project Minerva—and received twenty-eight negative comments and zero in support of it. "They are not developing them with understanding for vulnerability of suitability for sites."[22] A thirty-five-page report evaluating Project Minerva's potential impact on drinking water sources has twenty-one redacted pages—about 60 percent of the document—which somehow concludes it poses no threat to drinking water while redacting the methodology used to reach its conclusion.[23]

Any time the fossil fuel industry becomes part of a climate or environmental solution, people should be wary, says Karen Sokol. Up until the 2018 IPCC report, the industry was engaging in systematic and overt climate denial to undermine global climate goals. When the IPCC issued its report in 2018, the rhetoric shifted. Demands around the world for climate action were starting to grow, and they only got louder after that. Industry shifted its rhetoric to align itself with Paris climate goals and net zero. "They shifted from attacking any kind of climate solution to being the solution—being an essential part of the solution. Which at first blush make little sense. The climate crisis and rising emissions are by and large a result of fossil fuel production," says Sokol. [24]

PART OF THE SOLUTION?

Today the fossil fuel industry and its allies are "appropriating and weaponizing" language from climate advocates. ExxonMobil uses the phrase "advancing climate solutions" and "lower emission energy future." Shell is "working . . . to accelerate the transition to net-zero emissions." Chevron is "advancing a lower carbon future." All while planning billions of dollars in new oil and gas reserves. Some researchers have called this "fossil fuel solutionism." No longer are oil and gas executives straight-up "denying" that the climate is changing; instead, the message becomes one that ultimately slow-walks real climate action—saying, it's too expensive to address; it's too late to do anything.[25] "We call these 'climate delay' discourses, since they often lead to deadlock or a sense that there are intractable obstacles to taking action." Such discourses lead to deadlock by making obstacles to action against fossil fuels seem intractable.[26]

Oil companies will argue that the public wants energy, and the market dictates supply. But we can think of different solutions? Can we use less oil? "Do we need so many plastic forks and knives?" asked Verchick.[27] In places where natural gas is more expensive, like Europe, people are incentivized to use less of it. Can we rethink our own incentive structure? "We need to think about how the Louisiana delegation's support for carbon capture, Liquified Natural Gas (LNG) export terminals, and other heavy industrial output—not to mention, storing other people's carbon emissions—will make Black and Indigenous towns unlivable in Louisiana and around the world," said Verchick.[28]

Making the most aggressive adjustments through natural sinks, mass electrification, and conservation by less demand will be essential to avoiding the worst impacts of climate change. Achieving 1.5 degree Celsius over preindustrial levels compared to 2 degrees Celsius is the difference between saving 10 percent of all coral reefs or none at all. What we must understand is that we are headed for ruin, and we are now dibbling over the last 10 percent, said Sokol. The real question is, how will we adapt? We know decreasing production works. It not only reduces carbon dioxide, but other deadly air pollutants. We know restoring natural carbon sinks work. We know those should be prioritized. Anything else should be carefully and fully considered—not just this blanket all-of-the-above approach, but one that is thoughtful, says Sokol. Policy makers must think about how to get there, in addition to what the world will look like when we get there. "We are going to transition to another world one way or the other."[29]

THE SILVER LINING

The governor's Climate Action Plan—though imperfect—provides a documented account of the necessity of wetlands and mud for carbon sequestration. And despite the rosy, techno-utopianism of technologies like carbon capture, it projects quite urgently the need to maintain coastal wetlands and forests as important and effective carbon sinks. However, the logic of the working coast is explicitly entangled with the governor's plan. It's in the very first paragraph of the governor's executive order creating the Climate Action Task Force: "Whereas, Louisiana's working coast is a national treasure, exporting over $120 billion in annual goods, servicing 90 percent of the oil and gas activity in the Gulf of Mexico, producing 21 percent of all commercial fisheries landings by weight in the Lower 48 states, and providing winter habitat for five million migratory waterfowl . . . "[30]

The operationalization of wetlands for industrial output is a nefarious red herring. While acknowledging the importance of carbon sinks, it frames them through a calculus to merely achieve short-term economic gains. However, the existence of the record is important. The Climate Action Plan acknowledges the "business as usual approach" will not only raise overall greenhouse gas emissions but also lead to more coastal erosion, which also reduces natural carbon sinks. "Louisiana's abundant natural lands and wetlands are important not only for carbon sequestration but also for maintaining cultural heritage, coastal and agricultural economies, and reducing flood risk."[31]

And despite efforts to confuse and conflate the economic goals of capturing subsidized capital investments with the absolute environmental necessity of carbon reduction, the existence of this record will allow knowledge to escape the "plantation" of the working coast, as Anna Tsing might say. In the seepages of muddy spaces, we might nurture a knack for wildness that grows on its own volition despite efforts to control the narrative.

MUDDY THINKING AS A NECESSARY CONSTRAINT

Wildness, after all, challenges notions of linear progress. It is tactical and unplanned. It happens usually through neglect. It disrupts the colonizing impulse of space-time compression. In New Orleans and South Louisiana, I read wildness as the mud. As a concept, this idea of wild mud and Muddy Thinking writ large is meant to be broad and open to new and repurposed ideas, applications, and complexities. It is also speculative. We cannot abandon the mud no matter what sins of society it manages to contain.

We can embrace the constraints of the muddy thinking of subsistence epistemologies. And we must be willing to entertain unorthodox ideas—not just ideas, but new modes of living as well as *living with*. Just as mud itself is part of the working coast, so are the Indigenous residents, descendants of enslaved peoples, fishermen, subsistence farmers, and other agents of coproduction that Extractive Thinking fails to comprehend. We must embrace alliances to disrupt the Extractive Thinking that pervades the petro-economy. But I caution against the techno-utopian tendencies of what Lewis Mumford called the "technological sublime" for the next innovation to produce its way out of its own dilemma. Innovations in battery technology, for example, must also account for the creation of sacrifice zones of cadmium mining and labor conditions. We otherwise replace one extractive economy with another.

Here in New Orleans, sifting through the debris and uncollected city garbage left for weeks after Hurricane Ida, we faced that perennial question of recovery that arises in Louisiana somewhere every autumn: How to rebuild? What exactly are we building toward? We must be equipped to pose this question to every discipline, vocation, and person. Each year, we face not only technical preoccupations about better building codes and home elevations but also rising insurance premiums and social concerns like effects of trauma, closed schools, and community disruption, costs of recovery, and place-based identity. These questions span disciplines and require a strategy.

Such questions need to be communicated in ways that speak to the cultural foundations and subjectivities of a wide assortment of people and places. They must go to the heart of knowledge production. I'm thinking here of Toby Miller's proposal for a "green audit" of energy use for various cultural institutions and products; or Dipesh Chakrabarty's suggested syllabus for histories of energy that enabled the emergence of the Western enlightened subject. "Every evocation of Rousseau or Jefferson today needs to be accompanied by information on per capita energy use," he writes. A genealogy of energy use may denaturalize the historical epochs that were supported by extractive practices. A consideration of what energy humanities scholars call petrocultures might critique "the mansions of freedom" built on fossil fuels.[32] We may trace the contours not only of a political ecology of energy use and its environmental destruction but also the bourgeois culture and subjectivity it has produced.[33] The enlightened liberal subject moved from an ethos of stewardship of nature to domination and conquest through European colonialist ideology. "The idea of Nature being opposed to 'Man' privileges men in active relations with the natural world" while also feminizing nature.[34] Romanticism distanced people from the natural world, which was reinforced by urbanization and materialism. All of which led to a sentimentalization of wilderness. Rachel Carson so eloquently stated the still-unheeded call six decades ago: "Although modern man seldom remembers the fact, he could not exist without the plants that harness the sun's energy and manufacture the basic foodstuffs he depends upon for life."[35]

ENERGY HUMANITIES

What this means is that established academic scholarship regarding climate change in earth and climate sciences are enriched with the humanities and social sciences to study impacts of changing forces on local human and other-than-human ecologies. Such scholarship interrogates geopolitical decisions and engages with communities experiencing the local effects of those decisions. Rather than turn away, I suggest we look toward the Anthropocene as a concept providing for a potential opportunity for interdisciplinary exchange. To foster such exchange, we may follow models of care and "infrastructuring" to form alliances that cross disciplines to maintain critical spaces.[36]

We will need to accommodate multiple theoretical traditions: cultural analysis, political economy, poststructural analysis, environmental humanities, and science and technology studies. We will need these intellectual *and* affective tools to bring the "hyperobject" of climate change into full relief for citizens and policy makers to understand and intervene in equitable ways.[37] "Recognition of dying coral reef ecosystems in warming and acidifying seas was at the heart of advancing the very term *Anthropocene* in 2000." Coral taught biologists to understand the parochialism of their own ideas of individuals and collectives. "These critters taught people like me that we are all lichens, all coral," Haraway writes.[38]

(UN)NATURAL SCIENCE AND AN EQUITABLE LENS

Diverse practices will be needed to find possibilities in the rubble. While humankind has created "a species-level effect of a geologic force" (my apologies to the critics of the phrase), we have yet to establish a commensurate mode of collective thought and action to soften those effects.[39] Species-level responses will require an equity framework. Those who have profited most should contribute most. Ghosh argues that instead of one modernity, we should account for the effects of "plural modernities" of rising national powers in Western and Asian nations—as well as the price of foreclosing modernities to others. "What we have learned from this experiment is that the patterns of life that modernity engenders can only be practiced by a small minority of the world's population. . . . Every family in the world cannot have two cars, a washing machine, and a refrigerator—not because of technical or economic limitations but because humanity would asphyxiate in the process."[40] If we are serious about equity, we must be prepared to cushion the economic blow to those sidelined from the benefits of consumptive affluence and yet may still suffer from mandated reductions in consumption—as well as to those whose future modernities we are curtailing in the process.

How much money is it worth to the Global North, for instance, to place Brazil's Amazonian rain forest into an international trust? Recent moves by academic institutions to establish climate schools and interdisciplinary majors, as well as student-led protests against university endowments that invest in fossil fuel companies, signal popular interest. But we need a critical cultural approach to measure the effects of climate changes on its different and uneven sites of impact. We must interrogate the temptation of neoliberal techno-utopianism. Climate change is not isolated to only raising temperatures, but also production practices that are destroying habitats, polluting environmental justice communities, plasticizing oceans, and creating mass extinction. An equitable framework gives scholars, activists, artists, and community members the space to analyze climate and environmental discourses, including the abstraction of nature through mitigation, such as offsets and wetland mitigation banks that perversely encourage development in sensitive, irreplaceable areas.

Critics of the term "Anthropocene" argue that it lends itself to defeatism and cynicism, which allow for technocratic and geoengineering approaches. To encourage life, we need other forms of action and stories for inspiration. We need stories that affirm the need to resist the "stifling impotence created by the 'no possibility to do otherwise, whether we want it or not,' which now reigns everywhere."[41] The openly contested category of the Anthropocene offers a well-situated framework of intellectual exchange. It provides possibilities for cultural, intellectual, and critical conversations that elevate not only postcolonial concerns but also those of gender, race, affect, aesthetics, and other sociopolitical effects. Its utility comes from its widespread legibility among diverse practitioners. As a technoscientific category *and* a social critique, it in fact demands dialogues on social equity, government policy, postcolonial sensibility, and critical politics in the academy. I am also mindful of risks of critical erasure, which must be addressed openly.

The initial framing of the Anthropocene by geologists seemed to imply that "humanity" is responsible for catastrophic environmental change. But that has been indispensably critiqued. Not all humans are equally culpable. There is a great deal of important scholarship that questions the logics of attracting Anthropos embedded in the notion of the Anthropocene, says Haraway. Such critiques emphasize the uneven causes and consequences of global environmental change, as well as "the unmarked whiteness and Eurocentricity of Anthropocene discourses." Some social groups are more culpable than others for consumption, and others are overly burdened by its climatic effects. Marco Armiero uses the analogy of the *Titanic*, where poorer, steerage passengers experienced greater hardships than first-class passengers when the vessel broke in half and sank.[42]

The environmental justice veteran Robert Bullard is more explicit. People of color bear the brunt of environmental pollution because they are breathing other's people pollution. Black and Hispanic people are breathing much more pollution than they cause, and climate change will exacerbate these effects, he says. "Racism is a health issue. And racism is often times the underlying condition that creates, maintains, and underlines these disparities. I've been saying this for the last four decades or so."[43] Bullard's movement now is challenging and, more importantly, widening technical and economic conversations through social and community-level actions and making connections among different disciples and communities. The challenge is to do the necessary "care work" to keep those connections viable. That also means critiquing the hopeful yet vague executions of various international accords like the Paris Climate Agreement and the industry-led solutions at UN climate change conferences. The fairy tale of unending growth and consumption must be met head-on with a reckoning about the uneven wage effects that an economic drawdown might have on marginalized communities.[44]

An equitable lens, likewise, brings into conversations experiences of Indigenous communities, like tribes along the subsiding coast of Louisiana, who are also among those who have contributed the least to greenhouse gas emissions that drive climate change. "In fact, many Indigenous communities at risk of climate change impacts are the same ones that have already been—and continue to be—sacrificed by the fossil fuel extractive energy industry."[45] We must account for what the Native environmentalist and activist Winona LaDuke describes as "predator economics" that hold communities, and governments, hostage. The greatest negative impacts of our current extractive economic system fall on those places with few resources to resist. Health and justice must stay on the agenda of infrastructure investment. Ensuring a just future and a just transition will require difficult conversations among environmentalists as well. We must apply an equity lens on all green infrastructure project decisions, lest battery storage facilities create their own environmental justice communities. "There are huge facilities being planned next to black and brown neighborhoods without their consideration. We can be green, but we also have to be just," says Bullard. "Let's talk about siting issues."[46]

A critical Anthropocene, focusing on equity, can work with Muddy Thinking. We can slow down for difficult and important conversations. "Part of what going forward means to me is telling some really terrible stories about what's going on in the world," says Haraway.[47] Muddy Thinking invites the imperfection in these messy conjectures. It accepts the struggle in our unfolding moment. It provides a vocabulary with which to face the ruined places of the world to reaffirm our own humanity and allow some kind of healing to happen.

In late fall 2019, I attended a weeklong conference at Tulane University that was the culmination of a two-year project, titled "Mississippi: An Anthropocene River."[48] Activists, artists, scientists, scholars, and students had been meeting over two years along the river at various sites and presented their experiences at workshops in New Orleans. I was particularly riveted by a presentation from the Louisiana activist and artist Wendi Moore-O'Neal, who underscored the stakes of this kind of work. "Where you see along the river these petrochemical plants next to old plantations, many of the people there are descendants of people who worked those plantations," she said. The people of this place belong to the river, with all its complexities and contradictions. She continued, "You know there are so many things I don't know about the river, the industries, the environments . . . about birds, the trees, the land. . . . But some things I do know. I do know pieces of the histories of resistance to white supremacist, capitalist, imperialist, colonization. And I do know anywhere that that has existed, resistance to it has existed."

People of the river have managed to thrive and grow and build and create and love, she said. "That's what I want to share, is that there are people here. We're not simply beings to be researched—vehicles to master's degrees and PhDs. We are whole."

The ways of living and being here may look primitive, but they are important historically and for future survival, she continued. "We are the leaders of how we can survive being in resistance. . . . I'm talking about common, everyday living. The people on the street you are passing, whose names you do not know, know more about how to survive these constructs than any paper or any thesis or any doctorate, because they are here. Because we are here," she said. "We are still here. We are alive, and we are becoming."[49]

It was a powerful message about care—and carelessness. It was an intentional invitation for the artists, activists, and academics gathered over the week to trouble our own paradoxes.

There is after all a cost to producing knowledge. I'm thinking here of an essay that ran in *Southern Cultures* in 2021. Justin Hosbey and J. T. Roane wrote that while Katrina further stressed the swampy environments that had historically provided protection for maroon communities, it also solicited another kind of harm through academic efforts to understand the crisis. There's a neocolonialism embedded in certain attempts to understand environmental crises. Such framings tend to depict Black people either as equal partners in ecological destruction or as victims of extraction, they wrote. To counter the colonial binary, Hosbey and Roane set out to chart a maroon imaginary through cultural, spiritual, and intellectual lenses.[50] Their imaginary would find an ally in Muddy Thinking. A maroon imaginary might spend some time charting the cultural practices that pay homage to muddy entanglements, people, and places through New Orleans's Mardi Gras Indians. Enslaved people saw possibility in the mud because it gave masters the sense that there was "no place" for slaves to go. Their shared lineages are often not recorded or known. It is a culture that resists Extractive Thinking because it seems to resist origin itself. It expresses itself through call-and-response performance as well as exquisite craftsmanship in sewing unique outfits each Mardi Gras season. It is an ongoingness.

Taking up Moore-O'Neal's provocation—as well as the energy humanities approach—we might interrogate what kinds of knowledge we take without acknowledgment of the conditions that make it possible:

- A bus of professors touring neighborhoods struck by Katrina.
- The arborist taking a core sample through the trunk of a 500-year-old tree to date it.
- The WPA archaeologist slicing through a midden or burial mound to date the cultural landscape.
- The dystopian sea cruises taking thousands of passengers to witness the melting ice caps of Greenland.
- A grant to restore the Kid Ory Museum provided by the backers of the Wallace Grain Elevator.

It is a paradox: Our knowledge production is enabled by the very materials we critique and the people we overlook. I come away from this project—not trying to find a grand narrative, but instead gaining an appreciation for moments of disruption. And almost perversely, I find that the idea of permanence and posterity is part of the problem we now face. This ego-driven need to leave a lasting mark on the world has scarred it.

We've spent some time and attention on the extractive impulses that carved New Orleans out of the marshy swamp. But New Orleans subsists in the 2020s through a different kind of ethos. It is a damaged landscape, whose residents shape it through an imperfect practice of living. New Orleans is a harbinger of what happens after catastrophe. It demonstrates life in the ruins. Ironically, the Big Easy is anything but. It is a difficult place to live: boiling summer humidity; mosquitoes; crime; aging infrastructure; chopped streets; failing oversight; poor job prospects; subpar schools; routinely missed garbage pickup; uninspiring leadership; rising insurance costs; seasonal storms and evacuations; a perennial shortage of skilled labor. And perhaps most profoundly, there is an underlying racial enmity that culturally divides along a color line. In fact, as I write this, I'm wondering (perhaps with the reader), why live here? This is a good question.

Here's my response. New Orleans illustrates what James Scott calls a "thick city." New Orleans is beautiful and powerful in its decay. In its vulnerability, perhaps because of it, New Orleanians have developed an informal resiliency that comes with familiarity with death and catastrophe. As asserted recently in the *Gambit*'s observance of Black History Month, New Orleans was a place of both cruelty and hope. It was the birthplace of the first Black governor in the United States, PBS Pinchback, and the headquarters of Martin Luther King Jr.'s Southern Christian Leadership Conference. The city was home to Homer Plessy, the biracial shoemaker who was pulled from a first-class seat on a whites-only railcar en route to Covington in 1892. The act of civil disobedience, orchestrated by a New Orleans civil rights organization, violated Louisiana's 1890 Separate Car Act, which led to the eventual Supreme Court decision, *Plessy v. Ferguson*. The decision, as we know, legalized institutional racism throughout the South for nearly five decades until the *Brown v. Board of Education* decision in 1954.[51] It would be another six years for New Orleans itself to be integrated when four black students were permitted to attend the city's two whites-only public schools.[52] Here was also the largest domestic slave market—which increased the cruelty of that enterprise after the importation of slaves was banned by the United States in 1808. Egregious racism and the birth of the Civil Rights Movement, New Orleans is a place of contradictions that exist simultaneously. The city's history never quite retreats. It's archived in the mud.

This deep archaeology of culture is still influencing contemporary life. If we think of cities as a palimpsest of dreams and detritus, as Michael Sheringham suggests, we may gesture toward a place of possibility—which itself is a product of decomposition and archival traces.[53] New Orleans has been described as a simulacrum of itself, apparently frozen in time for the throngs of tourists looking for traces of its brothelized past in the gutted historic facades of French Quarter voodoo and daiquiri shops. But its aura emanates from a multilayered archive of street culture that complicates the racial binary of largely Protestant America. It was a place of slavery and free people of color; Indigenous tribes and European settlers; sensuality and American capitalism. "Neither mixed ancestry nor skin hue determined the status of freedom. . . . Some people of pure African ancestry acquired their freedom as some mulattoes remained enslaved."[54] Some enslaved people were hired out by their owners and were allowed to keep a percentage of their income, which was used for commerce or purchasing their own freedom.[55] New Orleans businessmen at times lobbied the city to allow for gatherings of the enslaved at Congo Square to whom they could sell merchandise. Slaveholders in English territories typically forbade people of African descent to assemble for traditional rituals or intermix with those from other places. "Yet in the French, Spanish, and American-ruled, Catholic-based city of New Orleans, African descendants—enslaved and free—perpetuated African Cultural practices and performances styles," which also included those Creole influences that developed in Cuba and Haiti and were embodied by enslaved people and free people of color who found their way to New Orleans.[56]

New Orleans was a place of fluidity—a site of disease, exoticism, and deception. The genealogies of the city and its people are traced around the footprints of its centuries of wayfarers. Today it's common for a single person to claim lineage to the oppressed and the oppressor.[57] These genealogies, unrecorded or deliberately forgotten, are nonetheless etched and written in the mud itself, cast in architectural cornices, and performed on public streets. They travel. They intermingle in art, music, and material culture. "The truth is you go down under Claiborne Avenue on a Sunday afternoon when there's a second line and look at the faces of people there—and they may identify as black but they are Atakapa, and they are Houma, and they are Chitimacha, and they are Choctaw," says Monique Verdin. "Go out anywhere in Louisiana, and you can see it. It's still there, and some may not be identifying as Indian-only, but we are still here and so much of what this place is and what would be."[58] New Orleans is a city where, according to the famed trumpeter Wynton Marsalis, "elegance met an indefinable wildness."[59] A romantic depiction notwithstanding, any description of New Orleans seems to slip along a palette of percussive rhythm, color, and sound. There is a wild and elegant substance that resides within the mud that constitutes the city—an improvised timelessness deposited by the meandering Mississippi. Its history remains on the surface to intermingle with the living. It offers an opening to rethink archaeology. Mud processes history, performing not an erasure but an embodiment of the past. It becomes the refuse and the bodies that are buried within it.

It demonstrates that there is life in futility, decay, and death. In a place of contradictions, it "exists under constant reminders of its contingent mortality."[60] Within this improbable place, there is something here to note—because it refuses the inevitable. The final two examples, which come from Rebecca Solnit and Rebecca Snedeker's inventive study of New Orleans called "Unfathomable City," illustrate the unique culture of this place that revels in the improbability of continuous rebirth. The first is the second line parade that follows the jazz funeral to ritualize and perform the celebratory affirmation of life. Joel Dinnerstein describes walking with the Prince of Wales Social Aid and Pleasure Club as they departed the small Spring Hill Missionary Baptist Church. Following the brass band, the second line of walkers adjust their stride with the changing rhythmic percussion as the energy and tempo shifts between human geography of present and past.

> After only two blocks, we slow the parade roll to honor the dead. The band downshifts into a dirge in front of the late Jimmy Parker's house on Annunciation and the Walers fall into a halting step with a syncopated slip: we strut in two lines with a slight diagonal step, shaping the air into chords of an ancient worship. Maybe we pick up his spirit. Maybe he's satisfied we are all still dancing for him. Once past Jimmy's house, the Tuba and snare drum pick up the groove, and down the block we pick up the Queen and her court.[61]

A joyous dance in the embrace of death. Decomposition enables life. The jazz funeral and the second line, the New Orleans street beat and the bamboula.[62] All represent a resistance to repression for people of color in New Orleans—which in 1856 ludicrously passed a brief city ordinance making it unlawful "to beat a drum, blow a horn or sound a trumpet in the city"[63]—and to the institutionalized racism of the Jim Crow South. Black New Orleanians responded with "social aid and pleasure clubs," holding not only social aid, but, more importantly, pleasure. This spirit is both memorialized in records and ritual and always being reinterpreted. In the face of interminable odds, it perseveres through a kind of baseline beat that it loves to play for itself. It is a human ecology of cultural bearers whose refusal requires and responds to its own improbability. A place of distinct and old neighborhoods, whose wards are mapped onto a byzantine coalition of political families, known by their acronyms: BOLD (in Central City), LIFE (in Mid City), COUP (in the Seventh Ward), and LAVA (in Lakeview).

Neighborhoods are mapped by their musical proclivities, whose province is as unique as the iconography and stitching of each costume of the two dozen Mardi Gras Indian tribes. My second example from Solnit and Snedeker's study comes from the Meters's bassist George Porter Jr., known as one of the progenitors of 1960s funk. Porter maps the geography of the city and his place in it through its musical styles. "In the studio, you know (Allen) Toussaint, he lays out a guideline about where we want to be. So Toussaint would say, 'We want to be in the Sixth Ward.' Or 'We going to the Ninth Ward.' Because if you went to different wards, you'd hear different styles of Music."[64]

The Lower Ninth Ward was gospel. The Upper Ninth Ward was blues. The Seventh and Sixth Wards were predominantly brass bands. "The Fifth . . . that was the police! Then, there was the Third Ward, where I grew up; that was the R&B guys, Earl King, Snooks [Eaglin]—that area was turning out those R&B musicians," Porter says. "And then there was Neville-ville. That was the Thirteenth." This foundation moves and shifts. Its low rhythm and bass give rise to the American sound of pop, rap, and rock and roll. Here at the end of the nation's river where everything collects. "And if they want to call us dirty—we say, yeah well it's your shit!" says Porter. "But we've taken their crap and made something wonderful out of it."[65]

Perhaps there is something to be learned here to cultivate care for one another and one's place. Perhaps there is something to be learned from hearing the joyous noise and then becoming it, in spite of yourself. New Orleans is always telling stories about itself—a practice that may have indeed saved it after Katrina and helped organize a national conversation around rebuilding in order to save its unique culture, a practice that might behoove all of us to start replicating. Stories about where we are now. There is possibility in decay. It is the birth of the imaginative soul, which may end up saving us.

INTRODUCTION: A TURN TO MUD

1. US Army Corps of Engineers, "Welcome to the J. Bennett Johnson Waterway."

2. Sherman was half brother to Gen. G. Mason Graham of Louisiana. During the war, he pardoned Louisiana governor Thomas Overton Moore and rendered aid to several cadets and professors captured in 1864—among them was Capt. David Boyd, who fought with the Louisiana Ninth Regiment. Boyd became superintendent after the war and is known for overseeing the relocation of the school to Baton Rouge. "William Tecumseh Sherman," LSU Military Museum; Hollandsworth, "The Burning of Alexandria."

3. "History of Alexandria, the Early Years," Alexandria-Louisiana.com.

4. Haraway, *Trouble*, 38.

5. Robertson, "We Built an App for That."

6. Giosan and Freeman, "How Deltas Work," 31.

7. Couvillion et al., "Land Area Change in Coastal Louisiana (1932 to 2016)."

8. Environmental Defense Fund, "Mapping Orphan Wells in the State"; Kang, "Workshop: Analyzing the Challenges of Improperly Abandoned and Orphaned Wells."

9. Schleifstein, "Orphaned Wells Increased by 50 Percent, Could Cost State Millions."

10. Peltz, comments to proposed rule regulating orphan wells, March 20, 2023.

11. Environmental Defense Fund, "Mapping Orphan Wells"; US GAO, "Offshore Oil and Gas."

12. Martin, Hollis, and Turner, "Effects of Oil-Contaminated Sediments on Submerged Vegetation."

13. Jacobs, "Monitoring and Enforcement of Air Quality"; Schleifstein, "Louisiana Should Better Identify Air Pollution Violations, Speed Enforcement."

14. Curole, in conversation with the author, 2016.

15. Haraway, *Trouble*, 4.

16. Nair et al., "Influence of Land Cover and Soil Moisture Based Brown Ocean Effect."

17. Pappas, "Why Did Hurricane Ida Stay So Strong for So Long?"

18. Dahl, "What More Gulf of Mexico Oil and Gas Leasing Means for Achieving U.S. Climate Targets" (testimony).

19. Intergovernmental Panel on Climate Change, "Sixth Assessment Report: Working Group 1."

20. Dahl, "What More Gulf of Mexico Oil and Gas Leasing Means for Achieving U.S. Climate Targets" (testimony).

21. Borenstein, "Carbon Dioxide Levels in Air Spike Past Milestone."

22. NOAA Research News, "Greenhouse Gas Pollution Trapped 49% More Heat in 2021 Than in 1990."

23. Hess, "Apocalypse When?" Emphasis in original.

24. Braun, "As Gov. Edwards Touts Greener Gas Alternatives at COP26, Activists Call It 'a Smokescreen.'"

25. Verchick, Harden, and Sokol, "The False Promise of Carbon Capture in Louisiana."

26. Howarth and Jacobson, "How Green Is Blue Hydrogen?"; Parker, "Blue Hydrogen Plant Touted for Louisiana."

27. Verchick, Harden, and Sokol, "The False Promise of Carbon Capture in Louisiana."

28. Szeman and Boyer, Introduction to *Energy Humanities*.

29. Ghosh, *Great Derangement*, 10.

30. Mukerji, *Modernity Reimagined*.

31. Szeman and Boyer, Introduction to *Energy Humanities*, 1.

32. Chakrabarty, "The Climate of History: Four Theses."

33. Ghosh, *Great Derangement*, 119.

34. Magelssen, *Simming*, 62–63.

35. Hughes, *Human-Built World*.

36. Ghosh, *Great Derangement*, 119.

37. Haraway, *Trouble*, 40–41.

38. Joselow, "Top Companies Are Undermining Their Climate Pledges with Political Donations."

39. Haraway, *Trouble*.

40. T. Morton, *Hyperobjects*.

41. Tsing, *Mushroom at the End of the World*.

42. Haraway, *Trouble*, 1.

43. Mitman, "Reflections on the Plantationocene."

44. "Trump: 'Why would I care about the climate?,'" *The Postillion*.

45. Carson, "Earth's Green Mantle."

46. Lyons, "Decomposition as Life Politics," 59.

47. Ibid.

48. Ibid., 73.

49. Keddy et al., "Wetlands of Lakes Pontchartrain and Maurepas," 54.

50. Day et al., *Perspectives on the Restoration*, 3.

51. Coastal surfaces in southeastern Louisiana "build down" as new vegetation springs up each year at a near-constant elevation. When vegetation cannot keep pace with this natural subsidence, it begins to submerge. Grasses die, and the area transitions to lakes linked by intertidal channels.

52. Keddy, "Wetlands of Lakes Pontchartrain and Maurepas."

53. Morris, *Big Muddy*.

54. Cowardin et al., "Classification of Wetlands and Deepwater Habitats."

55. US Energy Information Administration. "Louisiana State Energy Profile"; CPRA, "2017 Comprehensive Master Plan."

56. Shallat, "Holding Louisiana," 107.

57. Lin, "Louisianans Rally against Oil Drilling Moratorium."

58. Johnston, *Old River Control Structure, Louisiana: Hearing*.

59. Haraway, *Trouble*, 57.

60. M. Smith, "Louisiana Braces for New Flood Insurance Rates."

61. Associated Press, "Louisiana House Advances Bills to Address Insurance Crisis."

62. Baker, in conversation with the author, June 2, 2022.

63. US Army Corps of Engineers, "Notice of Intent to Prepare a Draft Environmental Impact Statement for the Lake Pontchartrain"; Smith and Schleifstein, "Corps Searches for Cause of Pump Corrosion."

64. Tsing, *Life in the Ruins*, 19.

1. A MUDSCAPE IN MOTION

1. Morris, *Big Muddy*.

2. Ibid., 10.

3. McPhee, "Atchafalaya."

4. Kniffen, Hiram, and Stokes, *Historic Indian Tribes of Louisiana*, 17.

5. Black, "Problem of the Mississippi."

6. Barry, *Rising Tide*, 38.

7. Schneider, *Old Man River*.

8. Shallat, "Holding Louisiana," 107.

9. The river's largest sediment flow probably came at the end of the nineteenth century as forestland in the Midwest was cleared for agriculture. Turner, "Mineral Sediment Loading."

10. Morris, *Big Muddy*, 2.

11. Twain, *Life on the Mississippi*, chapter 1.

12. Swanton, "Hernando de Soto's Route through Arkansas"; Whayne et al., "Spanish and French Explorations in the Mississippi Valley"; Kniffen, Hiram, and Stokes, *Historic Indian Tribes of Louisiana*.

13. Morris, *Big Muddy*, 12.

14. Ibid.

15. Childs and McNutt, "Hernando De Soto's Route from Chicaca."

16. Whayne, "Spanish and French Explorations," 34.

17. Morris, *Big Muddy*, 13.

18. Ibid., 12.

19. Kniffen, Hiram, and Stokes, *Historic Indian Tribes of Louisiana*, 4.

20. Ibid., 108–9, 115.

21. Dye, "Death March of Hernando de Soto."

22. Kniffen, Hiram, and Stokes, *Historic Indian Tribes of Louisiana*, 62.

23. Dye, "Death March of Hernando de Soto," 26–29, 31.

24. Kniffen, Hiram, and Stokes, *Historic Indian Tribes of Louisiana*, 5, 48.

25. Muth, "Once and Future Delta," 45.

26. Condrey, Hoffman, and Evers, "Last Naturally Active Delta Complexes."

27. Dye, "Death March of Hernando de Soto."

28. Muth, "Once and Future Delta."

29. Lyon, "New Deal Archaeology in the Southeast," 82; Morris, *Big Muddy*.

30. Kniffen, Hiram, and Stokes, *Historic Indian Tribes of Louisiana*, 18–21.

31. Ibid., 30.

32. Ibid., 18.

33. Ibid., 31.

34. Ibid.

35. Ibid., 108–9, 115.

36. Ibid., 3.

37. Ellwood et al., "LSU Campus Mounds."

38. Kniffen, Hiram, and Stokes, *Historic Indian Tribes of Louisiana*, 106.

39. Roberts, "Skull Fragment."

40. Kniffen, Hiram, and Stokes, *Historic Indian Tribes of Louisiana*, 122–23.

41. Ibid., 51; Ellis, *Great Power of Small Nations*.

42. Ibid., 107.

43. Ibid., 7.

44. Ibid.

45. Ibid.

46. Eggler, "FEMA Archeologists Find American Indian Pottery"; Brister, "Fort San Juan del Bayou."

47. F. W. Evans, *Congo Square*, 9.

48. Marx and Dawdy, "La Village des Chapitoulas."

49. Ibid.

50. Kniffen, Hiram, and Stokes, *Historic Indian Tribes of Louisiana*, 78, 96.

51. G. M. Hall, *Africans in Colonial in Louisiana*, 10.

52. Ibid.

53. Kniffen, Hiram, and Stokes, *Historic Indian Tribes of Louisiana*, 20.

54. Ibid.

55. G. M. Hall, *Africans in Colonial in Louisiana*, 8.

56. Day, *Perspectives on the Restoration of the Mississippi Delta*. Today Grand Isle is the only remaining inhabited barrier island.

57. Condrey, Hoffman, and Evers, "Last Naturally Active Delta Complexes"; Turner, "Mineral Sediment Loading."

58. Kniffen, Hiram, and Stokes, *Historic Indian Tribes of Louisiana*, 21.

59. Randolph, "River Activism."

60. Morris, *Big Muddy*.

61. Schneider, *Old Man River*, 329.

62. Ibid., 329.

63. The Lower Mississippi contains the so-called forever chemicals such as perfluorooctanesulfonic acid at more than 240 times current federal drinking water guidelines, as well as perfluorobutonic acid, in addition to concentrations of endrin, nitrogen, phosphorous,

nitrate nitrite benzene, carbon tetrachloride, hexachlorobenzene, poly-chlorinated biphe-nyls, styrene, arsenic, cadmium, lead, zinc, copper, mercury, and uranium. D. Mitchell, "'Forever Chemicals' Found."

64. McPhee, "Atchafalaya."

65. Ibid.

66. Bragg, *Historic Names and Places*, 195.

67. Ibid.

68. McPhee, "Atchafalaya."

69. Ibid.

70. These structures include the Old River Low Sill and Overbank Structures that began operation in 1962, the Old River Lock completed in 1963, and the Auxiliary Structure built in 1986. US Army Corps of Engineers, New Orleans District, "Old River Control."

71. T. Mitchell, *Rule of Experts*.

72. US Army Corps of Engineers, New Orleans District, "Old River Control."

73. Barra, "Good Sediment."

74. Together Louisiana, "Biggest Corporate Welfare Program in the Nation."

75. Scallan, "African American Cemeteries"; McDowell, "Sacred Ground."

76. F. W. Evans, *Congo Square*.

77. McConnaughey, "History in Bloom: Louisiana Irises, African Lilies May Mark Razed Black Community," April 3, 2022; "Fazendeville," Jean Lafitte National Historical Park and Preserve.

78. The Upper and Lower Nine were cleaved by the Industrial Canal, which opened in 1923 to link the Mississippi River and Lake Pontchartrain to compete against the Houston Ship Channel dug in 1914.

79. Schleifstein, "Dredging Mississippi."

80. R. Thompson, "Louisiana Ports Support More Than 396,000 Jobs."

81. In 1699, Bienville paddled out in two canoes to a small anchored English warship with ten cannons and dozens of settlers to inform them that France had already claimed the region with an established fortification upstream. The English captain, Louis Bond, believed the nineteen-year-old naval officer, turned around, and sailed away.

82. "Mississippi River—Ship Marine Traffic Live Tracking," Marinevesseltraffic.com.

83. "Mississippi River Sediment Delivery System," Louisiana Coastal Wetlands Conservation and Restoration Task Force.

84. Schleifstein, "'Power of the River.'"

85. "Hypoxia 101," Mississippi River/Gulf of Mexico Hypoxia Task Force, US EPA; Goolsby, "Mississippi Basin Nitrogen Flux"; Rabalais et al., "Characterization of Hypoxia."

86. Morris, *Big Muddy*.

2. MUDDY FOIL

1. Foucault, "Two Lectures" and "Truth and Power," 81.

2. Barry, *Rising Tide*, 421.

3. Klein and Zellmer, *Mississippi River Tragedies*, 46.

4. Smith, *How the Word Is Passed*, 6.

5. Leavitt, *Short History of New Orleans*, 136.

6. Independent Levee Investigation Team, "Geology of the New Orleans Region."

7. Colten, "Basin Street Blues."

8. Federal Writers' Project, *New Orleans City Guide*, 12.

9. Ibid., 17.

10. Leavitt, *Short History or New Orleans*, 85.

11. Dawdy et al., "Archaeological Investigations at St. Anthony's Garden," 7.

12. Lewis, *New Orleans: The Making of an Urban Landscape.*

13. Campanella, *Bienville's Dilemma.*

14. Between 1559 and 2008, an estimated 177 hurricanes have struck southern Louisiana. Rogers, "Development of the New Orleans Flood Protection System."

15. Campanella, "A Look Back at New Orleans' 300-Year-Long Drainage Drama."

16. F. W. Evans, *Congo Square*, 117.

17. Ibid., 44.

18. Campanella, "Whatever Became of the 'Lost Bayous.'"

19. Hosbey and Roane, "Totally Different Form of Living."

20. Federal Writers' Project, *New Orleans City Guide*, 12.

21. Pierson, *Tocqueville in America*, 622.

22. Federal Writers' Project, *New Orleans City Guide*, 156.

23. Ibid., 19.

24. Leavitt, *Short History of New Orleans*, 97.

25. Giblett, *Postmodern Wetlands*, 109.

26. Ibid.

27. D. Miller, *Dark Eden*, 48.

28. Giblett, *Postmodern Wetlands*, 109.

29. Ibid., 41.

30. Kelman, "Boundary Issues."

31. Giblett, *Postmodern Wetlands*, 119.

32. Ibid., 97.

33. Steiger, *Werewolf Book.*

34. Giblett, *Postmodern Wetlands*, 115.

35. Fields, "'This Is *Our* Land.'"

36. Hughes, *Human-Built World*, 43.

37. Le Miere, "Donald Trump Says 'Our Ancestors Tamed a Continent.'"

38. Giblett, *Postmodern Wetlands*, 217.

39. Ibid.

40. Ibid., 214.

41. Grant, "Deep in the Swamps."

42. G. W. Hall, *Africans in Colonial in Louisiana*, 202.

43. Ibid., 308.

44. Nevius, "New Histories of Marronage," 6.

45. Seck, *Bouki Fait Gombo*, 107.

46. Ibid., 142.

47. Ibid.; Follet, *Sugar Masters*, 266; "Slavery in Louisiana," Whitney Plantation Museum; Forensic Architecture, "Environmental Racism in Death Alley."

48. Seck, *Bouki Fait Gombo*, 108.

49. Aptheker, "Maroons within the Present Limits of the United States," 179; Diouf, "Slavery's Exiles."

50. Diouf, "Salvery's Exiles," 308.

51. Nevius, "New Histories of Marronage," 3.

52. Seck, *Bouki Fait Gombo*, 106.

53. Ibid., 107.

54. Nevius, "New Histories of Marronage," 14.

55. Aptheker, "Maroons within the Present Limits of the United States," 181.

56. Mathis and Weik, "Not Just Black and White."

57. Nevius, "New Histories of Marronage," 8.

58. T. C. Williams, *Self-Portrait in Black and White*, 28.

59. Ibid., 28.

60. C. Smith, *How the Word Is Passed*, 67.

61. Ibid., 68.

62. T. C. Williams, *Self-Portrait in Black and White*, 28.

63. C. Smith, *How the Word Is Passed*, 15.

64. Ibid., 15.

65. Ibid., 67.

66. Du Bois, "The Conservation of Races."

67. Forbes, "Mulattoes and People of Color in Anglo-North America."

68. T. C. Williams, *Self-Portrait in Black and White*, 21.

69. Mathis, "Reclaiming the Indigenous Past."

70. Forbes, "Mulattoes and People of Color in Anglo-North America," 20.

71. Forbes, "Mulattoes and People of Color in Anglo-North America," 52; see also Cheek, "Review of *Louisiana Creoles*," 578.

72. Kaplan-Levenson, "New Orleans: 300 // Bulchancha: 3000."

73. Mathis and Weik, "Not Just Black and White," 268.

74. Katz, *Black Indians*.

75. Katz, *Black Indians*; Healy, "Black, Native American and Fighting for Recognition."

76. Spears, "Freedmen Descendants."

77. Forbes, "Classification of Native Americans as Mulattoes."

78. Jeffery Darensbourg, speaking at Indigenous Day, organized by the Center for Gulf South at Tulane University, March 18, 2022.

79. Jessee, "Reshaping Louisiana's Coastal Frontier," 280.

80. Kniffen, Hiram, and Stokes, *Historic Indian Tribes of Louisiana*, 92–93.

81. Latour and Porter, *We Have Never Been Modern*, 10–11.

82. Leithart, "We Have Never Been Modern."

83. Forbes, "Mulattoes and People of Color in Anglo-North America," 23.

84. Chadwick, *Report on the Sanitary Condition*.

85. D. Miller, *Dark Eden*.

86. Foucault, *Security, Territory, Population*, 19.

87. Ibid., 21.

88. Colten, "Basin Street Blues."

89. Campanella, "Disaster and Response."

90. Southmayd, *Report of the Howard Association of New Orleans.*

91. Campanella, "A Look Back at New Orleans' 300-Year-Long Drainage Drama."

92. Kelman, "Boundary Issues," 699.

93. Bauer, "Yellow Fever."

94. Roach, *Cities of the Dead.*

95. Kelman, "Boundary Issues," 699.

96. Ibid.; Barry, *Rising Tide.*

97. Campanella, "Above-Sea-Level New Orleans."

98. Federal Writers' Project, *New Orleans City Guide,* 14.

99. E. J. Blum, "Crucible of Disease."

100. Ellis, *Yellow Fever and Public Health,* 84.

101. Colten, "Basin Street Blues," 30.

102. Bauer, "Yellow Fever"; Watts, "Yellow Fever, Malaria and Development," 216.

103. Bauer, "Yellow Fever," 366.

104. Colten, "Basin Street Blues," 243.

105. Campanella, "A Look Back at New Orleans' 300-Year-Long Drainage Drama."

106. Kelman, "Boundary Issues."

107. D. W. Davis, "Historical Perspective," 89.

108. Colten, "Basin Street Blues."

109. Ibid., 250.

110. Rogers, "Development of the New Orleans Flood Protection System."

111. Gagliano, "Canals, Dredging, and Land Reclamation."

112. Campanella, "A Look Back at New Orleans' 300-Year-Long Drainage Drama."

113. Ibid.

114. Campanella, "Disaster and Response," 21.

115. Campanella, "A Look Back at New Orleans' 300-Year-Long Drainage Drama."

116. During heavy rainstorms, the city's streets fill in areas that are above sea level. In one investigation, it was discovered that several of the water pumps had malfunctioned, likely from the stress caused by congestion of debris. The CEO of the sewerage and water board resigned. When forensic teams went through the system in 2018, they pulled out 46 tons of Mardi Gras beads. B. Evans, "Forty-Six Tons of Mardi Gras Beads."

117. Shallat, "Holding Louisiana," 103.

118. S. J. Jackson, "Building the Virtual River."

119. Tripp and Herz, "Wetland Preservation and Restoration."

120. Lynn Greenwalt, foreword to Cowardin et al., "Classification of Wetlands and Deepwater Habitats."

121. Cowardin et al., "Classification of Wetlands and Deepwater Habitats," 3.

122. Ibid., iii.

123. Escobar, "Construction Nature."

124. This is slightly different from other areas like the Florida Everglades, where arguably the wetlands' value became linked to the need for municipal drinking water for the sizable population of South Florida.

125. CPRA, "2017 Comprehensive Master Plan for a Sustainable Coast," ES-10.

3. THE VILLAINOUS RIVER

1. Murphy, "Maneuvers on the Mississippi River."

2. Grabar, "Hell Is High Water."

3. Ibid.

4. McConnaughey, "History in Bloom"; "Fazendeville," Jean Lafitte National Historical Park and Preserve.

5. Burke, *New Iberia Blues*, 2.

6. Calder and Wilkinson, "Mississippi River Treacherous."

7. Morris, "Only a River."

8. Twain, *Life on the Mississippi*, chapter 13.

9. Ibid.

10. How, *James B. Eads*, 7.

11. Barry, *Rising Tide*, 38.

12. Twain, *Life on the Mississippi*, chapter 10.

13. Ibid.

14. Sandlin, *Wicked River*, 9.

15. Ibid., 9.

16. Ibid., 22.

17. Ibid., 23–24.

18. Klein and Zellmer, *Mississippi River Tragedies*.

19. How, *James B. Eads*.

20. Barry, *Rising Tide*, 26.

21. How, *James B. Eads*, 6.

22. Ibid.

23. McPhee, "Atchafalaya."

24. How, *James B. Eads*, 7.

25. Twain, *Life on the Mississippi*.

26. Sandlin, *Wicked River*, xxi.

27. Ibid., 20.

28. Ibid., 21.

29. Ibid.

30. Ibid., 15.

31. Ibid., 16.

32. Ibid., 13.

33. Ibid., 28.

34. Ibid., 27.

35. Bragg, *Historic Names and Places*, 396.

36. Ibid.

37. Ibid., 193.

38. Ibid.

39. Ibid., 200.

40. Ibid., 205.

41. Sandlin, *Wicked River*, 9.

42. Ibid., 226.

43. Layton, "Mirror-Twins."

44. For every action, there is an equal and opposite reaction.

45. O'Neill, *Rivers by Design*.

46. Shallat, *Structures in the Stream*, 4.

47. Ibid.

48. Ibid., 154.

49. Ibid., 154–56.

50. Scott, *Seeing Like a State*, 83.

51. Ibid.

52. Lippincott, "History of River Improvement," 636.

53. Randolph, "River Activism."

54. Gibbons v. Ogden, 22 U.S. 1 (1824).

55. Pabis, "Delaying the Deluge."

56. Campanella, "How River Diversions Powered and Fed Early New Orleans."

57. Reuss, "Army Corps of Engineers and Flood-Control Politics."

58. Reuss, "Art of Scientific Precision," 297.

59. Ibid., 293.

60. Ibid.

61. Ibid.

62. Ibid.

63. Reuss, "Andrew A. Humphreys."

64. Rogers, "Development of the New Orleans Flood Protection System," 604.

65. Wright, *Swamp and Overflowed Lands in the United States*; O'Neill, *Rivers by Design*, 49.

66. Barry, *Rising Tide*, 34.

67. O'Neill, *Rivers by Design*, 40.

68. Ibid., 51.

69. A. R. Hall, "Public Slaves and State Engineers."

70. Jessee, "Reshaping Louisiana's Coastal Frontier."

71. Rothman, "Georgetown University and the Business of Slavery."

72. Dattel, "Cotton in a Global Economy."

73. Harrison, *Levee Districts and Levee Building*, 13.

74. Dattel, "Cotton in a Global Economy."

75. Gudmestad, "Steamboats," 393.

76. Ibid.

77. Ibid., 394.

78. Ibid., 395.

79. A. R. Hall, "Public Slaves and State Engineers," 564.

80. Ibid., 535.

81. Ibid.

82. Ibid., 569.

83. Ibid., 546.

84. Ibid., 548–49.

85. Ibid., 532.

86. Ibid., 534, 542.

87. Ibid., 547.

88. Ibid., 570.

89. Ibid.

90. Gudmestad, "Steamboats," 392.

91. Ibid., 408.

92. Kniffen, Hiram, and Stokes, *Historic Indian Tribes of Louisiana*, 85.

93. Ibid., 88.

94. Ibid., 91.

95. Barry, *Rising Tide*, 34–35.

96. Reuss, "Andrew A. Humphreys."

97. Ibid.

98. Pabis, "Delaying the Deluge."

99. Ibid.

100. Shallat, "Building Waterways," 41.

101. Pabis, "Delaying the Deluge," 439.

102. US Army Corps of Engineers, "Andrew Atkinson Humphreys (1810–1883)."

103. Reuss, "Andrew A. Humrpheys," 3.

104. Pabis, "Delaying the Deluge," 424.

105. Shallat, *Structures in the Stream*, 206.

106. Ibid.

107. Barry, *Rising Tide*.

108. Ibid., 75.

109. Ibid., 82.

110. Ibid.

111. O'Neill, *Rivers by Design*.

112. Lippincott, "History of River Improvement."

113. Reuss, "Andrew A. Humphreys," 24.

114. Pabis, "Delaying the Deluge," 453.

115. Reuss, "Andrew A. Humphreys," 11.

116. Ibid., 33.

117. Pearcy, "Ransdell-Humphreys Flood Control Act."

118. Saikku, "Taming the Rivers."

119. Barry, *Rising Tide*, 358.

120. Saikku, "Taming the Rivers."

121. Manders, "US Army Corps of Engineers and the Mississippi River Cutoff Plan."

4. THE BIRTH OF RIVER SCIENCE
AND GRASSROOTS GREENWASHING

1. Geology was formally established at LSU during the period from 1892 to 1898 and designated the Department of Geology, Mineralogy and Botany. By 1922, the department had been renamed the Department of Geology.

2. Louisiana State University, "Brief History of LSU Geology and Geophysics."

3. Ibid.

4. Anderson, *Richard Joel Russell*.

5. Ibid.

6. Louisiana State University, "Brief History of LSU Geology and Geophysics."

7. Anderson, *Richard Joel Russell*, 251.

8. Ibid., 376.

9. Lyon, *A New Deal for Southeastern Archaeology*, 81.

10. Ibid., 82.

11. Ibid., 81.

12. The start of World War II was the beginning of the end of WPA archaeology as government funds were steered toward the war effort.

13. Rees, *Archaeology of Louisiana*.

14. Anderson, *Richard Joel Russell*, 377.

15. Ibid., 376.

16. Arnold, *A Thousand Ways*.

17. Anderson, *Richard Joel Russell*, 377.

18. Louisiana State University, "Brief History of LSU Geology and Geophysics."

19. Anderson, *Richard Joel Russell*, 378–79.

20. Ibid., 376.

21. Ibid., 381.

22. Reed and Wilson, "Coast 2050," 17.

23. Mathewson and Shoemaker, "Louisiana State University Geography at Seventy-Five."

24. Fisk, "Geological Investigation of the Alluvial Valley."

25. Coleman and Gagliano, "Sedimentary Structures."

26. Robinson, "Harold N. Fisk."

27. LeBlanc, "Harold Norman Fisk," 22; Day et al., *Perspectives on the Restoration*.

28. LeBlanc, "Harold Norman Fisk," 22.

29. Ibid.; Randolph, "Modeling Authority over a Drowning Coast."

30. Kniffen, Hiram, and Stokes, *Historic Indian Tribes of Louisiana*, 41.

31. Gagliano, Meyer-Arendt, and Wicker, "Land Loss," 1684.

32. Gagliano, "Canals, Dredging," 23.

33. Ibid., 30.

34. Gagliano, Light, and Becker, "Controlled Diversions."

35. Wicker, "Mississippi Deltaic Plain."

36. Theriot, *American Energy*.

37. Gagliano, "Canals, Dredging," 1.

38. Arnold, *A Thousand Ways*, Kindle Location 543.

39. About 80 percent of Louisiana's coastal lands are privately owned. Wilkins et al., "Preliminary Options for Establishing Recreational Servitudes."

40. D. W. Davis, "Louisiana Canals and Their Influence on Wetland Development."

41. Ibid.

42. Arnold, *A Thousand Ways*, Kindle Location 111.

43. Ibid.; D. W. Davis, "Louisiana Canals and Their Influence on Wetland Development."

44. Houck, "Reckoning," 190; Arnold, *A Thousand Ways*.

45. Houck, "Reckoning," 193.

46. Ibid.

47. Ibid.

48. Arnold, *A Thousand Ways*, Kindle Location 1189.

49. Ibid.

50. Ibid., Kindle Location 1059.

51. Houck, "Reckoning," 193.

52. Theriot, *American Energy*, 149.

53. Ibid., 96.

54. Houck, "Reckoning," 194.

55. Theriot, *American Energy*, 151.

56. Houck, "Reckoning," 205.

57. The federal agency was established in 1982 to manage the development of the petroleum resources in the deep waters.

58. Johnston, Cahoon, and La Peyre, "Technical Summary for 'Outer Continental Shelf.'"

59. Neill and Turner, "Backfilling Canals."

60. Baustian and Turner, "Restoration Success of Backfilling."

61. Houck, "Reckoning," 206.

62. Ibid.

63. The CPRA annual plan for fiscal year 2023 was $1.35 billion, 80 percent of which went toward restoration construction projects. CPRA, "Fiscal Year 2023 Annual Plan."

64. Bahr, "A Little Scrutiny," 3.

65. Houck, "Reckoning," 207.

66. Turner, in conversation with the author, 2017.

67. Ibid.

68. Day et al., "Life Cycle of Oil and Gas Fields," 12.

69. Baustian and Turner, "Restoration Success of Backfilling."

70. Equal to three-fourths of the circumference of the Earth.

71. Turner and McClenachan, "Reversing Wetland Death."

72. Ibid.

73. Morton, Buster, and Krohn, "Subsurface Controls on Historical Subsidence Rates," 767.

74. Houck, "Reckoning," 220.

75. Morton and Bernier, "Recent Subsidence-Rate Reductions," 559.

76. "Subsidence," City of Long Beach.

77. Arnold, *A Thousand Ways*.

78. Houck, "Reckoning," 9.

79. Ibid., 10.

80. Houck, phone conversation with the author, 2017.

81. Houck, phone conversation with the author, 2017.

82. Reed, phone conversation with the author, 2018.

83. Houck, "Reckoning," 195.

84. Reed and Wilson, "Coast 2050: A New Approach to Restoration," 8.

85. Ibid.

86. Kolker, Allison, and Hameed, "An Evaluation of Subsidence Rates," L.21404.

87. Houck, "Reckoning," 196.

88. Day et al., "Life Cycle of Oil and Gas Fields," 12.

89. Arnold, *A Thousand Ways*.

90. Day et al., "Life Cycle of Oil and Gas Fields," 12.

91. Persaud, "Louisiana EarthQuakes."

92. McLindon, "Oil and Gas Geologist."

93. Day et al., "Life Cycle of Oil and Gas Fields."

94. Theriot, *American Energy*, 133.

95. Ibid.

96. Ibid., 135.

97. Ibid.; Houck, "Reckoning."

98. Houck, "Reckoning," 255.

99. Theriot, *American Energy*; Houck, "Reckoning"; Coalition to Restore Coastal Louisiana, "Coastal Louisiana, Here Today and Gone Tomorrow?," 1.

100. Coalition to Restore Coastal Louisiana, "Coastal Louisiana, Here Today and Gone Tomorrow?," 17–18.

101. Houck, "Reckoning," 211; Coalition to Restore Coastal Louisiana, "Coastal Louisiana, Here Today and Gone Tomorrow?," 16.

102. Coalition to Restore Coastal Louisiana, "Coastal Louisiana, Here Today and Gone Tomorrow?"

103. Ibid.

104. Theriot, *American Energy*, 144.

105. Houck, "Reckoning," 211.

106. Theriot, *American Energy*, 144.

107. Houck, "Reckoning," 187.

108. Ibid., 211.

109. Ibid., 212.

110. Theriot, *American Energy*, 162.

111. Louisiana Coastal Wetlands Conservation and Restoration Task Force, "Louisiana Coastal Wetlands Restoration Plan."

112. CWPPRA, "What Is CWPPRA?"

113. Reed and Wilson, "Coast 2050."

114. Ibid., 12.

115. However, the CWPPRA, or Breaux Act, continues to fund restoration projects.

116. Boesch et al., "Scientific Assessment of Coastal Wetland Loss."

117. Reed and Wilson, "Coast 2050," 5.

118. Louisiana Coastal Wetlands Conservation and Restoration Task Force and the Wetlands Conservation and Restoration Authority, "Coast 2050," 40.

119. Sanders, "Blanco v. Burton." Until 1937, the Department of the Interior deferred mineral lease activity to states, which it recognized as owners of submerged lands. But Roosevelt's interior secretary, Harold Ickes, announced that the federal government—not the states—would grant oil leases offshore. The Truman administration extended federal jurisdiction over the "the natural resources of the subsoil and seabed of the continental shelf beneath the high seas."

120. Sanders, "Blanco v. Burton," 259, 263.

121. Turner and McClenachan, "Reversing Wetland Death."

122. Theriot, *American Energy*, 187.

123. Sanders, "Blanco v. Burton"; Theriot, *American Energy*, 187.

124. Theriot, *American Energy*, 187.

125. Ibid., 188.

126. Ibid., 189.

127. Caffey and Schexnayder, "Coastal Louisiana and South Florida."

128. Ibid., 5.

129. Often wetland "loss" was intentional. So-called swamp-buster programs launched in the nineteenth century were actively supported by Louisiana and Florida among other states as a mechanism to transform wetlands into cultivatable land.

130. Caffey and Schexnayder, "Coastal Louisiana and South Florida."

131. Ibid.

132. Ibid.

133. Ibid.; emphasis original.

134. Houck, "Reckoning," 213.

135. Milling's uncle had also formed Louisiana Land & Exploration, an enormous oil producer and once one of the largest private landowners in South Louisiana. Another client was Continental Land and Fur Co., whose original interests in muskrat pelts had long ago been replaced by its land's mineral rights. Houck, "Reckoning."

136. Houck, "Reckoning," 213.

137. Theriot, *American Energy*, 192.

138. Houck, "Reckoning," 213.

139. America's WETLAND Foundation, press releases archives.

140. Ibid.

141. Houck, "Reckoning."

142. America's WETLAND Foundation, press releases archives.

143. Houck, "Reckoning." America's WETLAND Foundation's major donor list identified Shell as its "World Sponsor." Sustainability Sponsors include Chevron, ConocoPhillips, and ExxonMobil. America's WETLAND Foundation sponsored an ad in the journal article, Caffey and Schexnayder, "Coastal Louisiana and South Florida."

144. America's WETLAND Foundation, "America's Vanishing Wetland National Campaign Launched to Save Coastal Louisiana," press releases archive.

145. Ibid.

146. Theriot, *American Energy*, 195.

5. THE "KATRINA EFFECT" AND THE WORKING COAST

1. Theriot, *American Energy*, 197.

2. Louisiana Recovery Authority, "Progress Report." Louisiana is the largest producer of shrimp, oyster, and blue crab in the nation.

3. CPRA, "2007 Comprehensive Master Plan."

4. Ibid.

5. Klein, *Shock Doctrine*, 4; C. Johnson, *Neoliberal Deluge*.

6. "Domenici Plans New Drive on OCS," *Inside Energy Extra*, 1.

7. Amaewhule, "Hurricane Katrina Renews Push for Drilling."

8. Hebert, "Katrina Spurs New Debate."

9. Helman, "Open the Spigots."

10. Shields, "Bill Ties La. Aid, ANWR Drilling."

11. J. Blum, "Offshore Drilling Backers Smell Victory."

12. Louisiana Recovery Authority, "Progress Report."

13. The passage of GOMESA created a model for future administrations to incentivize congressional support for drilling in states protected by drilling moratoriums off the shores of Virginia, North Carolina, South Carolina, and Alaska.

14. Bureau of Ocean Energy Management, "Gulf of Mexico Energy Security Act."

15. CPRA, "2017 Comprehensive Master Plan"; Environmental Defense Fund, "Groups Pleased as Key Sediment Diversions Advance."

16. CPRA, "2007 Comprehensive Master Plan."

17. Isaacson, "Greatest Education Lab."

18. Klein, *Shock Doctrine*, 4.

19. Mallach, "Where Will People Live?"

20. Campanella, *Delta Urbanism*.

21. Ibid.

22. Freudenburg, "Disproportionality and Disaster"; Rivlin, "Why New Orleans's Black Residents Are Still Under Water after Katrina."

23. The Louisiana Recovery Authority focused its lobbying efforts on the following: two additional block grants in June 2006 and November 2007 totaling $7.2 billion; appropriations of $7.1 billion for levee repairs and a commitment from the Bush administration to fund another $7.6 billion to complete the $14.5 billion levee system around metropolitan New Orleans; and passing the 2007 Water Resources Development Act, overriding a veto by President Bush, that authorized nearly $7 billion for hurricane protection projects in Louisiana, with $2 billion in funding for coastal restoration. Louisiana Recovery Authority, "Progress Report."

24. Ibid., 4.

25. Southeast Louisiana Flood Protection Authority–East (Flood Protection Authority) website.

26. Act 8, Senate Bill No. 71 (duplicate of House Bill No. 141), First Extraordinary Session of the Louisiana Legislature (2005), 21.

27. Act 8, Senate Bill No. 71 (2005).

28. CPRA, "2007 Comprehensive Master Plan."

29. Lopez, "Multiple Lines of Defense Strategy to Sustain Coastal Louisiana," 187.

30. CPRA, "2007 Comprehensive Master Plan."

31. Ibid., 30.

32. Direct land loss in Louisiana threatens $3.4 billion in assets, $7.4 billion in economic activity, and up to 12,200 jobs. A storm striking the New Orleans area in a future with increased land loss could result in a $133 billion increase in storm damages. Economic disruptions from this storm event could affect 26,000 establishments, 320,000 employees, and $50 billion in output above the impacts of a similar event occurring on today's coast. CPRA, "2007 Comprehensive Master Plan."

33. Ibid., 29.

34. Swyngedouw, Kaka, and Castro, "Urban Water."

35. CPRA, "2017 Comprehensive Master Plan"; Schleifstein, "Louisiana Granted Final Funds."

36. Festa, "Race against Time."

37. CPRA, "2007 Comprehensive Master Plan."

38. Brown, *Undoing the Demos*, 115–18.

39. CPRA, "2007 Comprehensive Master Plan," 27.

40. Touting itself as the steward for the important Mississippi River dates to the nineteenth century as an argument for federal support for navigation and flood control.

41. A 2015 study conducted by LSU and the Rand Corporation estimated that at least $100 billion of energy and petroleum infrastructure is at risk from a receding coastline over the next 25 years, including annual sales of $2.4 to $3.1 billion and associated payroll between $400 and $575 million; and between $5.8 billion and $7.4 billion in annual "output." Similarly, they estimated that increased storm damage could have a total impact on the nation of between $8.7 and $51.5 billion and increased disruption to economic activity leading to $5 to $51 billion in total lost output, including indirect and induced effects. CPRA, "2012 Comprehensive Master Plan."

42. Ibid., 121.

43. CPRA, "Deep Horizon Oil Spill Restoration"; CPRA, "Multiyear Implementation and Expenditure Plan."

44. CPRA, "Deep Horizon Oil Spill Restoration."

45. The National Fish and Wildlife Foundation (NFWF) is in charge of funding projects using the criminal penalties. Sack and Schwartz, "Left to Louisiana's Tides."

46. A study showed an increase of 2 degrees Celsius would raise the water surrounding the city levees by 14.5 inches by 2040 and 6.5 feet by 2100. A separate study by the Coastal Geologic Survey predicted that changes in precipitation and temperature would have unforeseen changes in the marsh coastal vegetation regardless of sea level rise, which will have effects on the coast.

47. Haselle, "Voluntary Relocation."

48. M. Smith, "New Data Shows Sharp Flood Insurance Hikes."

49. Ravits, "'A Long Way to Go.'"

50. Hammer, "What Is Causing Louisiana's Insurance Crisis."

51. CPRA, "Clarity on GOMESA Funding and FY19 Annual Plan Presented at CPRA Board Meeting," December 13, 2017, http://coastal.la.gov/wp-content/uploads/2017/12/2017.12.13-GOMESA.pdf.

52. Gagliano, "Canals, Dredging."

53. CPRA, "2017 Comprehensive Master Plan," E5–11.

54. Ibid.

55. Ibid., 157.

56. Water Campus website.

57. Martin, "Contracting for Louisiana Coastal Restoration."

58. Legendre, "Coastal Plan Could Help Diversify Economy."

59. Brown, *Undoing the Demos*, 116.

60. Ibid.

61. Big River Coalition website.

62. Turner, "Discussion of: Olea, R. A. and Coleman, J. L., Jr., 2014"; Penland et al., *Process Classification of Coastal Land Loss between 1932 and 1990*.

63. Jackson and Chapple, *Breakpoint*.

64. Houck, "Reckoning."

65. M. Davis, in conversation with the author, September 24, 2018.

66. Moertle, telephone interview by the author, August 31, 2018.

67. Reed, phone conversation with the author, 2018.

68. Ibid.

69. Reed and Wilson, "Coast 2050"; Morton, Bernier, and Barras, "Evidence of Regional Subsidence"; CPRA, "2012 Comprehensive Master Plan," Appendix A1: Projects Screened Out of Consideration in the 2012 Coastal Master Plan, accessed May 17, 2023, https://coastal.la.gov/our-plan/2012-coastal-masterplan/cmp-appendices/.

70. "Subsidence," City of Long Beach.

71. Turner, "Doubt and the Values of an Ignorance-Based World View for Restoration"; Turner, "Mineral Sediment Loading."

72. Colten and Hammering, "Social Impact Assessment Methodology," 61–66.

73. Thorington, "Mid-Barataria Sediment Diversion Project."

74. Avenal v. State, 03-C-3521 (La. 10/19/04), 886 So. 2d 1085.

75. Sneath, "Louisiana Hopes New Oyster Leases Will Ease Pain."

76. Bradberry, "Coastal Protection Guest Column: Louisiana Land Loss a Crisis."

77. Ricks, in conversation with the author, 2016.

78. Knowles, "Four More Dead Dolphins Wash Ashore"; Louisiana Office of the Governor, "Gov. Edwards Requests Federal Disaster Declaration."

79. Ricks, public comments observed by the author, Coastal Protection and Restoration Board Meeting, 2019.

80. Hunter et al., "Using Natural Wetlands for Municipal Effluent Assimilation."

81. "Hypoxia 101," Mississippi River/Gulf of Mexico Hypoxia Task Force, US EPA; Goolsby, "Mississippi Basin Nitrogen Flux"; Rabalais et al., "Characterization of Hypoxia."

82. Roberts, "Deadline in Plaquemines Parish–State Standoff."

83. Colten, "The Place for Humans in Louisiana Coastal Restoration."

84. Sack and Schwartz, "Left to Louisiana's Tides."

85. Jackson and Chapple, *Breakpoint*.

86. Ibid., 59.

87. Ricks, in conversation with the author, 2016.

88. Kearney, Rider, and Turner, "Freshwater River Diversions for Marsh Restoration."

89. Maldonado, "Multiple Knowledge Approach for Adaptation to Environmental Change."

90. T. Miller, *Greenwashing Culture*, 74.

91. J. Williams, "Louisiana Coastal Wetlands."

92. T. Miller, *Greenwashing Culture*, 79.

93. Archer, public comments observed by the author, Communities Adaptation Leadership Forum, August 29, 2018.

94. Ambrose, "BHP Billiton Backs BP's Return to Gulf."

95. Bartlett, "Foreign Bribery Update."

96. Yeomans and Bowater, "One Year On, Brazil Battles to Rebuild."

97. Laville, "BHP Reveals Five Mine Dams at 'Extreme' Risk."

98. M. Davis, in conversation with the author, September 24, 2018.

99. Maldonado, "Multiple Knowledge Approach for Adaptation to Environmental Change."

100. The disconnect is not lost on the fishermen who live downstream of New Orleans. "Why are we the ones that got to adapt?!" yelled one of them during a heated town hall that was organized by the CPRA and Louisiana Sea Grant to discuss ways to adapt to the changes that diversions will be bringing to the marshes.

INTERLUDE: A MODEST PROPOSAL

1. Rivlin, *Katrina: After the Flood.*

2. Elie, "Why We Came Home."

3. D. E. Taylor, *Toxic Communities.*

6. MUD, PLASTICS, AND CANCER ALLEY

1. Baurick, "No Cleanup Planned."

2. McPhee, "Atchafalaya."

3. Lerner, *Diamond*, 9.

4. Ibid.

5. Ibid.; Baurick, Younes, and Meiners, "Welcome to 'Cancer Alley.'"

6. Boyle, "'A Slam upon Our State.'"

7. Schleifstein, "EPA Investigates Louisiana Environmental, Health Agencies."

8. Allen, "Popular Geography of Illness," 193.

9. Ibid.

10. Together Louisiana, "Biggest Corporate Welfare Program."

11. Center for International Environmental Law, "Fueling Plastics."

12. Carter, "Inside Exxon's Playbook."

13. "Louisiana Climate Action Plan: Climate Initiatives Task Force Recommendations to the Governor," February 2022.

14. McMichael, "Plant Location Factors," 24.

15. Peterson, *Giants on the River*, 10.

16. McMichael, "Plant Location Factors," 24.

17. Allen, "Popular Geography of Illness," 178.

18. Leber, "Your Plastic Addiction."

19. Peterson, *Giants on the River*, 10.

20. Ibid.

21. McMichael, "Plant Location Factors," 40.

22. Ibid.

23. H. Davis, *Plastic Matter*, 3.

24. Ibid.

25. Allen, "Popular Geography of Illness," 179.

26. B. Williams, "'That we may live.'"

27. Allen, "Popular Geography of Illness."

28. Houck, "Shintech: Environmental Justice," 472.

29. Forensic Architecture, "Environmental Racism in Death Alley," 6.

30. Houck, "Shintech: Environmental Justice," 459.

31. Davies, "Slow Violence and Toxic Geographies," 417.

32. Peterson, *Giants on the River*, 8.

33. Tahir, in conversation with the author, April 2022.

34. Smith, *How the Word Is Passed*; see also Seck, *Bouki Fait Gombo*.

35. Forensic Architecture, "Environmental Racism in Death Alley."

36. Coastal Environments, Inc., "Cartographic Regression Analysis"; G. W. Hall, *Africans in Colonial Louisiana*; Seck, *Bouki Fait Gombo*.

37. Forensic Architecture, "Environmental Racism in Death Alley," 58; Follet, *Sugar Masters*.

38. Forensic Architecture, "Environmental Racism in Death Alley," 58; see also Follett, *Sugar Masters*, 111.

39. Muhammad, "The Sugar that Saturates the American Diet"; "The 1619 Project."

40. Forensic Architecture, "Environmental Racism in Death Alley."

41. Ibid., 53.

42. Vaughan, "Louisiana Sugar," 95.

43. Lindner, "Sugar Granulation on the Boré Plantation"; "Sugar at LSU: A Chronology," exhibition.

44. Vaughn, "Louisiana Sugar," 95.

45. State Library of Louisiana Historic Photograph Collection, "Sugar Kettle outside of the chemical engineering building at Louisiana State University."

46. Austin, "Coastal Exploitation, Land Loss and Hurricanes," 675.

47. G. W. Hall, *Africans in Colonial Louisiana*.

48. G. W. Hall, *Africans in Colonial Louisiana*, 346.

49. Fiehrer, "Baron de Carondelet," 473.

50. Faber, "Slave Insurrections in Louisiana."

51. Kniffen, Hiram, and Stokes, *Historic Indian Tribes of Louisiana*, 64.

52. G. W. Hall, *Africans in Colonial Louisiana*, 344.

53. Ibid., 345.

54. Bradshaw, "Saint-Domingue Revolution."

55. Ibid.

56. C. Smith, *How the Word Is Passed*, 53.

57. S. Thompson, "No Sweetness Is Light," 72.

58. Vaughan, "Louisiana Sugar."

59. Seck, *Bouki Fait Gombo*, 115.

60. Bradshaw, "Saint-Domingue Revolution"; Vaughan, "Louisiana Sugar."

61. Vaughan, "Louisiana Sugar," 97.

62. Bradshaw, "Saint-Domingue Revolution."

63. C. Smith, *How the Word Is Passed*, 53.

64. Faber, "Slave Insurrections in Louisiana."

65. Buman, "To Kill Whites."

66. Faber, "Slave Insurrections in Louisiana."

67. Ibid.

68. Buman, "To Kill Whites," 78.

69. Ibid., 73.

70. Ibid., 81.

71. Ibid., 19.

72. C. Smith, *How the Word Is Passed*; Seck, *Bouki Fait Gombo*, 113.

73. Buman, "To Kill Whites," 13.

74. C. Smith, *How the Word Is Passed*, 55.

75. Ibid.

76. Thompson, "No Sweetness Is Light," 72.

77. Seck, *Bouki Fait Gombo*, 114.

78. Buman, "To Kill Whites," 14.

79. Ibid., 79.

80. C. Smith, *How the Word Is Passed*, 55.

81. Buman, "To Kill Whites," 82.

82. C. Smith, *How the Word Is Passed*, 64.

83. Ibid.

84. Ibid., 54.

85. Faber, "Slave Insurrections in Louisiana."

86. C. Smith, *How the Word Is Passed*, 54.

87. Ibid., 65.

88. Forensic Architecture, "Environmental Racism in Death Alley," 2.

89. Laurent, "History of St. John the Baptist Parish," 82.

90. Louisiana Bucket Brigade, "Down by the River."

91. Du Bois, "Freedman's Bureau."

92. Ibid.

93. Davies, "Slow Violence and Toxic Geographies."

94. Ibid.

95. Forensic Architecture, "Environmental Racism in Death Alley," 4.

96. Lerner, *Sacrifice Zones.*

97. Forensic Architecture, "Environmental Racism in Death Alley," 4.

98. Kardas-Nelson, "The Petrochemical Industry Is Killing Another Black Community"; US EPA, "National Air Toxics Assessment; Request for Reconsideration."

99. Schleifstein, "EPA Investigates Louisiana Environmental, Health Agencies." The EPA's National Air Toxics Assessments in 2015 listed chloroprene as a likely carcinogen and should be limited to .2 microgram per cubic meter to comply with the limit of acceptable risk of 100 in one million people. O'Donnell, "Management Alert: Prompt Action Needed."

100. Schleifstein, "EPA Investigates Louisiana Environmental, Health Agencies."

101. Ibid.

102. Lerner, *Diamond*, 3; Lerner, *Sacrifice Zones.*

103. Bullard and Wright, "The Politics of Pollution," 71.

104. Downey and Hawkins, "Race, Income, and Environmental Inequality."

105. Fleischman and Franklin, *Fumes across the Fence-Line.*

106. Bullard et al., "Toxic Wastes and Race at Twenty," 176.

107. Lerner, *Sacrifice Zones.*

108. Tessum et al., "Inequity in Consumption of Goods and Services."

109. Lerner, *Sacrifice Zones*.

110. Blodgett, "Analysis of Pollution and Community Advocacy in 'Cancer Alley.'"

111. Terrell and St Julien, "Air Pollution Is Linked to Higher Cancer Rates."

112. Schleifstein, "Louisiana Should Better Identify Air Pollution Violations."

113. Jacobs, "Monitoring and Enforcement of Air Quality."

114. Baurick, Younes, and Meiners, "Welcome to 'Cancer Alley'"; Terrell and James, "Racial Disparities in Air Pollution Burden and COVID-19 Deaths"; Flatt, "Gasping for Breath."

115. Baurick, Younes, and Meiners, "Welcome to 'Cancer Alley.'"

116. Jacobs, "Monitoring and Enforcement of Air Quality"; Schleifstein, "Louisiana Should Better Identify Air Pollution Violations."

117. Meiners, "Polluters Paradise."

118. Bracket, "Oil, Chemical Plants Released Tons of Pollutants"; Sneath, "'Ticking Time Bombs': Residents Kept in the Dark."

119. US Energy Information Administration, "Louisiana State Energy Profile."

120. Natter, "Ida Leaves Toxic Chemicals."

121. Sneath, "Louisiana Shell Refinery Left Spewing Chemicals."

122. Sneath, "Louisiana Shell Refinery Left Spewing Chemicals."

123. Together Louisiana, "Biggest Corporate Welfare Program."

124. Tahir, in conversation with the author, April 2022.

125. Lavigne, "Formosa Plastic Toxic Tour."

126. Ibid.

127. Coastal Environments, Inc., "Cartographic Regression Analysis"; Center for Biological Diversity, Healthy Gulf, Louisiana Bucket Brigade, Rise St. James v. Army Corps et al. (2020).

128. Forensic Architecture, "Environmental Racism in Death Alley."

129. Gobert, Rolfes, and Kray, "Genealogy Zoom Call."

130. Forensic Architecture, "Environmental Racism in Death Alley," 73.

131. Ibid.

132. D. Mitchell, "Judge Tosses Air Permit for Giant Plastics Complex."

133. Simmonds, Lavigne, and Rolfes, "Army Corps Orders Full Environmental Review."

134. Gobert, Rolfes, and Kray, "Genealogy Zoom Call," July 7, 2021.

135. Ibid.

136. Ibid.

137. Forensic Architecture, "Environmental Racism in Death Alley," 69.

138. US Energy Information Administration, "Louisiana State Energy Profile," 2018.

139. Gobert, Rolfes, and Kray, "Genealogy Zoom Call," July 7, 2021.

CONCLUSION: STUCK IN THE MUD

1. Haraway, *Trouble*, 37.

2. McKenna, "IPCC Understated the Need to Cut Emissions from Methane"; Intergovernmental Panel on Climate Change, "Sixth Assessment Report: Working Group 1."

3. Shindell, "Global Methane Assessment"; Ripple et al., "World Scientists' Warning."

4. Haraway, *Trouble*, 37.

5. Tsing, *Mushroom*.

6. Haraway, *Trouble*, 37; Tsing, *Mushroom*.

7. Haraway, *Trouble*, 37.

8. Ibid.

9. Smethurst, "Nature Writing," 30.

10. Haraway, *Trouble*.

11. Abdulla et al., "Explaining Successful and Failed Investments in U.S. Carbon Capture and Storage Using Empirical and Expert Assessments," *Environmental Research* (2021), Lett. 16 014036, https://iopscience.iop.org/article/10.1088/1748-9326/abd19e/meta.

12. Verchick, Harden, and Sokol, "False Promise of Carbon Capture."

13. Ibid.; Carbon Capture Coalition, accessed May 18, 2023, https://carboncapturecoalition.org/wp-content/uploads/2023/04/CCC_federalpolicyblueprint_2023.pdf.

14. Verchick, Harden, and Sokol, "False Promise of Carbon Capture."

15. Ibid.; "500+ Organizations Call on US and Canadian Leaders," Waterkeeper Alliance.

16. "Louisiana Climate Action Plan: Climate Initiatives Task Force."

17. Verchick, Harden, and Sokol, "False Promise of Carbon Capture."

18. "Louisiana Climate Action Plan: Climate Initiatives Task Force."

19. DeJong et al., "Consequence Analysis of the Draft Portfolio of Climate Strategies and Actions."

20. Hilburn, "How Louisiana Became the Carbon Capture Capital of the South."

21. Ibid.

22. Ibid.

23. Sneath, "Oil and Gas Industry Is Using Louisiana's Climate Task Force"; "Hydrology Report for Project Minerva."

24. Verchick, Harden, and Sokol, "False Promise of Carbon Capture."

25. Levantesi, "Climate Deniers and the Language of Climate Obstruction."

26. Lamb et al., "Discourses of Climate Delay."

27. Verchick, Harden, and Sokol, "False Promise of Carbon Capture."

28. Ibid.

29. Ibid.

30. State of Louisiana Executive Order Number JBE2020-18, Climate Initiatives Task Force.

31. "Louisiana Climate Action Plan: Climate Initiatives Task Force."

32. Chakrabarty, "Climate of History."

33. Swyngedouw, Kaka, and Castro, "Urban Water."

34. Smethurst, "Nature Writing," 30.

35. Carson, "Earth's Green Mantle."

36. Striphas, "Caring for Cultural Studies"; Pezzullo, "Resisting Carelessness."

37. T. Morton, *Hyperobjects*; "Hyperobjects, Hyposubjects and Solidarity," *Anthropocenes—Human, Inhuman, Posthuman*.

38. Haraway, *Trouble*, 72.

39. Chakrabarty, "Climate of History."

40. Ghosh, *Great Derangement*, 91.

41. Haraway, *Trouble*, 128.

42. Armiero, "Filtering the Anthropocene."

43. Bullard, "Quest for Environmental and Climate Justice."

44. Thunberg, "'How Dare You' Speech."

45. Maldonado, *Seeking Justice in an Energy Sacrifice Zone*, 10.

46. Bullard, "Quest for Environmental and Climate Justice."

47. Mitman, "Reflections on the Plantationocene."

48. Sponsored by the Max Planck Institute, Haus der Kulturen der Welt (HKW), and New Orleans Center for the Gulf South at Tulane University and ByWater Institute.

49. Moore-O'Neal, speaking at the Anthropocene River Campus: The Human Delta at Tulane University, November 2019.

50. Hosbey and Roane, "A Totally Different Form of Living."

51. Rifkin, "Overlooked No More."

52. Fazio, "'McDonogh Three' Integrated New Orleans Schools."

53. Sheringham, "Archiving."

54. F. W. Evans, *Congo Square*, 50.

55. Ibid., 38, 113.

56. Ibid., 115.

57. Kaplan-Levenson, "New Orleans: 300 // Bulchancha: 3000."

58. Ibid.

59. Marsalis, "Saving America's Soul Kitchen."

60. Solnit and Snedeker, *Unfathomable City*, 83.

61. Ibid., 110.

62. F. W. Evans, *Congo Square*, 37.

63. Ibid., 31.

64. Solnit and Snedeker, *Unfathomable City*, 119.

65. Ibid.

Abdulla, Ahmed, et al. "Explaining Successful and Failed Investments in U.S. Carbon Capture and Storage Using Empirical and Expert Assessments." *Environmental Research* (2021), Lett. 16 014036. https://iopscience.iop.org/article/10.1088/1748-9326/abd19e/meta.

Act 8, Senate Bill No. 71 (duplicate of House Bill No. 141). First Extraordinary Session of the Louisiana Legislature (2005), 21. Accessed May 17, 2023. http://www.columbia.edu/itc /journalism/cases/katrina/State%20of%20Louisiana/Louisiana%20State%20Legislature /LA%20Senate/SB71%20Coastal%20Protection.pdf.

Allen, B. "Popular Geography of Illness in the Industrial Corridor." In *Transforming New Orleans and Its Environs*, edited by Craig Colten, 178–201. Pittsburgh: University of Pittsburgh Press, 2001. http://www.jstor.org/stable/j.ctt7zw9kz.

———. *Uneasy Alchemy: Citizens and Experts in Louisiana's Chemical Corridor Disputes.* Cambridge, MA: MIT Press, 2003.

Amaewhule, Olivia. "Hurricane Katrina Renews Push for Drilling in Restricted Offshore Areas." *IHS Global Insight*, September 12, 2005.

America's WETLAND Foundation. Press releases. Way Back Machine. Accessed May 15, 2023. https://web.archive.org/web/20200808122016/www.americaswetland.com/category /press-releases/.

Ambrose, Jillian. "BHP Billiton Backs BP's Return to Gulf of Mexico with £1.75bn Investment." *Telegraph*, February 9, 2017. Accessed May 17, 2023. https://www.telegraph.co.uk /business/2017/02/09/bhp-billiton-backs-bps-return-gulf-mexico-175bn-investment/.

Anderson, Charles. *Richard Joel Russell, 1895–1971: A Biographical Memoir.* Washington, DC: National Academy of Sciences, 1975. http://www.nasonline.org/publications /biographical-memoirs/memoir-pdfs/russell-richard-j.pdf.

Aptheker, Herbert. "Maroons within the Present Limits of the United States." *Journal of Negro History* 24, no. 2 (April 1939). https://doi.org/10.2307/2714447.

Archer, Rachel. Public comments observed by the author. Communities Adaptation Leadership Forum, Nicholls State University, Thibodaux, LA, August 29, 2018.

Armiero, Marco. "Filtering the Anthropocene." The Anthropocene Project: A Report. The Anthropocene Campus, November 14, 2014. https://www.youtube.com/watch?v =9gabtRVTvCI.

Arnold, John T. *A Thousand Ways Denied: The Environmental Legacy of Oil in Louisiana.* Natural World of the Gulf South. Baton Rouge: LSU Press, 2020.

Associated Press. "Louisiana House Advances Bills to Address Insurance Crisis." *Biz New Orleans,* February 2, 2023. Accessed May 16, 2023. https://www.bizneworleans.com /louisiana-house-advances-bills-to-address-insurance-crisis/.

Austin, Diane E. "Coastal Exploitation, Land Loss, and Hurricanes: A Recipe for Disaster." *American Anthropologist* 108, no. 4 (December 2006): 671–91. https://doi.org/10.1525 /aa.2006.108.4.671.

Avenal v. State, 03-C-3521 (La. 10/19/04), 886 So. 2d 1085. Accessed May 16, 2023. http:// www.lasc.org/opinions/2004/03c3521.opn.pdf.

Bahr, Len. "A Little Public Scrutiny Shows LA's Public Trust Being Screwed by Shrewd Energy Interests." LaCoastPost.com, May 15, 2017. https://web.archive.org/web/20150404032049 /http://lacoastpost.com/blog/.

Baker, David. In conversation with the author. Studio in the Woods, New Orleans, June 2, 2022.

Barra, Monica P. "Good Sediment: Race and Restoration in Coastal Louisiana." *Annals of the American Association of Geographers* 111, no. 1 (2021): 266–82. https://doi.org/10.1080 /24694452.2020.1766411.

Barry, John. *Rising Tide: The Great Mississippi Flood of 1927 and How It Changed America.* New York: Simon & Schuster, 1997.

Bartlett, Audin. "Foreign Bribery Update: A Harsh Lesson for a Global Miner." Carter Newell Lawyers, June 2015. Accessed May 16, 2023. https://www.carternewell.com/icms_docs /218542_Foreign_bribery_update_A_harsh_lesson_for_a_global_miner.pdf.

Bauer, J. H. "Yellow Fever." *Public Health Reports (1896–1970)* 55, no. 9 (March 1, 1940): 362–71. https://doi.org/10.2307/4583195.

Baurick, Tristin. "No Cleanup Planned as Millions of Plastic Pellets Wash Up along Mississippi River and Flow to the Gulf." *New Orleans Advocate Times-Picayune,* August 18, 2020. Accessed May 16, 2023. https://www.nola.com/news/environment/article_b4fba760 -e18d-11ea-9b0b-b3a2123cf48b.html.

Baurick, Tristin, Leah Younes, and Joan Meiners. "Welcome to 'Cancer Alley,' Where Toxic Air Is about to Get Worse." *ProPublica* and *New Orleans Advocate Times-Picayune,* October 30, 2019. Accessed May 16, 2023. https://www.propublica.org/article/welcome -to-cancer-alley-where-toxic-air-is-about-to-get-worse.

Baustian, Joseph J., and R. Eugene Turner. "Restoration Success of Backfilling Canals in Coastal Louisiana Marshes." *Restoration Ecology* 14, no. 4 (December 2006): 636–44.

The Big River Coalition. Louisiana Maritime Association. Accessed May 12, 2023. http:// www.bigrivercoalition.org/.

Black, W. M. "The Problem of the Mississippi." *North American Review* 224, no. 838 (December 1927): 630–43. https://www.jstor.org/stable/25110389.

Blodgett, A. D. "An Analysis of Pollution and Community Advocacy in 'Cancer Alley': Setting an Example for the Environmental Justice Movement in St James Parish, Louisiana." *Local Environment* 11, no. 6 (2006): 647–61. https://doi.org/10.1080/13549830600853700.

Blum, Edward J. "The Crucible of Disease: Trauma, Memory, and National Reconciliation during the Yellow Fever Epidemic of 1878." *Journal of Southern History* 69, no. 4 (November 2003): 799–826. https://doi.org/10.2307/30040097.

Blum, Justin. "Offshore Drilling Backers Smell Victory." *Washington Post*, February 21, 2006, final edition. Accessed May 17, 2023. https://www.washingtonpost.com/wp-dyn/content/article/2006/02/20/AR2006022001380.html.

Boesch, Donald, et al. "Scientific Assessment of Coastal Wetland Loss, Restoration and Management in Louisiana." Special issue, *Journal of Coastal Research*, no. 20 (1994): i–v, 1–103.

Borenstein, Seth. "Carbon Dioxide Levels in Air Spike Past Milestone." Associated Press, June 4, 2022. Accessed May 16, 2023. https://apnews.com/article/climate-science-national-oceanic-and-atmospheric-administration-environment-oceans-24753128b5e1 1aca8d69de4072380798.

Boyle, Louis. "'A slam upon our state': Republican Senator Takes Offense to Biden's Remarks on Louisiana's 'Cancer Alley.'" *Independent*, February 4, 2021. Accessed September 1, 2023. https://www.independent.co.uk/climate-change/bill-cassidy-biden-louisiana-cancer-alley-environment-pollution-b1797627.html.

Bracket, R. "Oil, Chemical Plants Released Tons of Pollutants While Shutting Down for Hurricane Laura." Weather.com, August 31, 2020. Accessed June 4, 2021. https://weather.com/news/news/2020-08-31-hurricane-laura-pollution-refineries-chemical-plants.

Bradberry, Johnny. "Coastal Protection Guest Column: Louisiana Land Loss a Crisis; Here's What We're Doing about It." *Advocate* (Baton Rouge), April 3, 2018. Accessed May 16, 2023. https://www.theadvocate.com/baton_rouge/opinion/article_7179c8c2-3691-11e8-a12a-63644815819e.html.

Bradshaw, Jim. "Saint-Domingue Revolution." 64 Parishes, October 22, 2014. Accessed May 17, 2023. https://64parishes.org/entry/saint-domingue-revolution.

Bragg, Marion. *Historic Names and Places on the Lower Mississippi River*. Vicksburg, MS: US Army Corps of Engineers, 1977.

Braun, Paul. "As Gov. Edwards Touts Greener Gas Alternatives at COP26, Activists Call It 'a Smokescreen.'" WRKF.com, November 2, 2021. Accessed May17, 2023. https://www.wrkf.org/news/2021-11-02/as-gov-edwards-touts-greener-gas-alternatives-at-cop26-activists-call-it-a-smokescreen.

Brister, Nancy. "Fort San Juan del Bayou Photos and History of the 'Old Spanish Fort' Bayou St. John at Lake Pontchartrain New Orleans, Louisiana." Accessed May 12, 2023. http://old-new-orleans.com/NO_SanJuan_del_Bayou.html.

Brown, Wendy. *Undoing the Demos: Neoliberalism's Stealth Revolution*. New York: Zone Books, 2017.

Bullard, R. "Quest for Environmental and Climate Justice: The Continuing Story of John Muir and the American Environmental Movement." In *Sierra Club: 125 Years of Protecting Nature*, edited by Tom Turner, 93–103. San Francisco: Sierra Club Books, 2017.

Bullard, R., M. Paul, R. Saha, and B. Wright. "Toxic Wastes and Race at Twenty." *Race*, March 2007, 176.

Bullard, R. D., and B. H. Wright. "The Politics of Pollution: Implications for the Black Community." *Phylon* 47, no. 1 (1986): 71–83. https://doi.org/10.2307/274696.

Buman, Nathan. "To Kill Whites: The 1811 Louisiana Slave Insurrection." Master's thesis, Louisiana State University, 2008. Accessed May 12, 2023. https://digitalcommons.lsu.edu/gradschool_theses/1888.

Bureau of Ocean Energy Management. "Gulf of Mexico Energy Security Act (GOMESA)." Accessed July 27, 2018. https://www.boem.gov/Oil-and-Gas-Energy-Program/Energy-Economics/Revenue-Sharing/Index.aspx.

Burke, James Lee. *The New Iberia Blues: A Dave Robicheaux Novel.* London: Orion Books, 2019.

Caffey, R. H., and M. Schexnayder. "Coastal Louisiana and South Florida: A Comparative Wetland Inventory." NSGL Document #LSU-G-03-021, National Sea Grant Library, 2003. Accessed May 16, 2023. http://eos.ucs.uri.edu/seagrant_Linked_Documents/lsu/lsug03021.pdf.

Calder, Chad, and Missy Wilkinson. "Mississippi River Treacherous for Accident Victims, Difficult for Recovery Efforts, Experts Say." NOLA.com, April 28, 2022. Accessed May 17, 2023. https://www.nola.com/news/mississippi-river-treacherous-for-accident-victims-difficult-for-recovery-efforts-experts-say/article_c4453df6-c5a4-11ec-9248-bb8a17de9eb6.html.

Campanella, Richard. "Above-Sea-Level New Orleans: The Residential Capacity of Orleans Parish's Higher Ground." Edited by Douglas J. Meffert et al. Center for Bioenvironmental Research, April 2007. Accessed May 17, 2023. http://richcampanella.com/assets/pdf/study_Campanella%20analysis%20on%20Above-Sea-Level%20New%20Orleans.pdf.

———. *Bienville's Dilemma: A Historical Geography of New Orleans.* Lafayette: Center for Louisiana Studies, University of Louisiana at Lafayette, 2008.

———. *Delta Urbanism: New Orleans.* Chicago: Planners Press of the American Planners Association, 2010.

———. "Disaster and Response in an Experiment Called New Orleans, 1700s–2000s." In *Oxford Research Encyclopedia of Natural Hazard Science*, March 2016. https://dx.doi.org/10.1093/acrefore/9780199389407.013.1.

———. "How River Diversions Powered and Fed Early New Orleans, Sometimes amid Controversy." *New Orleans Advocate Times-Picayune*, June 1, 2022. Accessed May 17, 2023. https://www.nola.com/entertainment_life/campanella-how-river-diversions-powered-and-fed-early-new-orleans-sometimes-amid-controversy/article_96fe2408-e0e2-11ec-89cb-8faa7fa9e843.html.

———. "A Look Back at New Orleans' 300-Year-Long Drainage Drama." *New Orleans Advocate Times-Picayune*, August 22, 2018. Accessed May 17, 2023. https://www.nola.com/archive/a-look-back-at-new-orleans-300-year-long-drainage-drama/article_1b472e73-e021-5d3b-bfaa-450bfdb56940.html.

———. "Whatever Became of the 'Lost Bayous' of New Orleans? Waterways Once Laced the Metro Area." *New Orleans Advocate Times-Picayune*, June 30, 2021. Accessed July 7, 2021. https://www.nola.com/entertainment_life/whatever-became-of-the-lost-bayous-of-new-orleans-waterways-once-laced-the-metro-area/article_ee9ad300-d8f0-11eb-93fb-434d4af30cbc.html.

Carbon Capture Coalition. Carbon Capture Coalition. Accessed May 18, 2023. https://carboncapturecoalition.org/wp-content/uploads/2023/04/CCC_federalpolicyblueprint_2023.pdf.

Carson, Rachel. "Earth's Green Mantle." In *Silent Spring*, Anniversary ed., 267–94. Boston: Houghton Mifflin, 2002. Originally published 1962.

Carter, Lawrence. "Inside Exxon's Playbook: How America's Biggest Oil Company Continues to Oppose Action on Climate Change." *Unearthed*, June 3, 2021. Accessed May 17, 2023. https://unearthed.greenpeace.org/2021/06/30/exxon-climate-change-undercover/.

Center for Biological Diversity, Healthy Gulf, Louisiana Bucket Brigade, Rise St. James v. Army Corps et al. Case 1:20-cv-00103 (2020). Accessed May 18, 2023. https://www.biologicaldiversity.org/campaigns/plastic-production/pdfs/2020_01_14-Formosa-404-Complaint.pdf.

Center for International Environmental Law. "Fueling Plastics: How Fracked Gas, Cheap Oil, and Unburnable Coal Are Driving the Plastics Boom." 2017. Accessed May 17, 2023. https://www.ciel.org/wp-content/uploads/2017/09/Fueling-Plastics-How-Fracked-Gas-Cheap-Oil-and-Unburnable-Coal-are-Driving-the-Plastics-Boom.pdf.

Chadwick, E. *Report on the Sanitary Condition of the Labouring Population of Great Britain* . . . London: Printed by W. Clowes and Sons for Her Majesty's Stationery Office, 1843.

Chakrabarty, Dipesh. "The Climate of History: Four Theses." *Critical Inquiry* 35, no. 2 (2009): 197–222. https://doi.org/10.1086/596640.

Cheek, Gary C., Jr. "Review of *Louisiana Creoles: Cultural Recovery and Mixed-Race Native American Identity*." *American Indian Quarterly* 33, no. 4 (2009): 577–79.

Childs, H. Terry, and Charles H. McNutt. "Hernando De Soto's Route from Chicaca through Northeast Arkansas: A Suggestion." *Southeastern Archaeology* 28, no. 2 (2009): 165–83. http://www.jstor.org/stable/40713517.

Coalition to Restore Coastal Louisiana. "Coastal Louisiana, Here Today and Gone Tomorrow? A Citizens' Program for Saving the Mississippi River Delta Region to Protect Its Heritage, Economy and Environment: Draft for Public Review." April 1987. Accessed May 17, 2023. https://www.govinfo.gov/app/details/CZIC-qh541-5-c65-c63-1987.

Coastal Environments, Inc. "Cartographic Regression Analysis of Certain Tracts of Land Located in T.11S and T.12S., R.15E. (Southeastern Land District West of the Mississippi River)." St. James Parish, Louisiana, February 19, 2020.

Coastal Protection and Restoration Authority (CPRA). "Clarity on GOMESA Funding and FY19 Annual Plan Presented at CPRA Board Meeting." December 13, 2017. Accessed May 17, 2023. http://coastal.la.gov/wp-content/uploads/2017/12/2017.12.13-GOMESA.pdf.

———. "Deep Horizon Oil Spill Restoration: RESTORE Act." Accessed May 17, 2023. http://coastal.la.gov/deepwater-horizon-oil-spill-content/oil-spill-overview/restore-act/.

———. "Fiscal Year 2023 Annual Plan." State of Louisiana, Baton Rouge. Accessed May 16, 2023. https://ap23.coastal.la.gov/.

———. "Multiyear Implementation and Expenditure Plan." Baton Rouge, 2015. Accessed May 17, 2023. http://coastal.la.gov/wp-content/uploads/2015/05/Draft-RESTORE-Act-Multiyear-Implmentation-and-Expenditure-Plan.pdf.

————. "2007 Comprehensive Master Plan for a Sustainable Coast." Office of the Governor, Baton Rouge, 2007.

————. "2012 Comprehensive Master Plan for a Sustainable Coast." Office of the Governor, Baton Rouge, 2012.

————. "2017 Comprehensive Master Plan for a Sustainable Coast." Office of the Governor, Baton Rouge, 2017.

Coastal Wetland Planning, Preservation, and Restoration Act (CWPPRA). "What Is CWP-PRA?" Accessed June 14, 2019. https://lacoast.gov/new/Default.aspx.

Coleman, James, and Sherwood Gagliano. "Sedimentary Structures: Mississippi River Deltaic Plain." In *Primary Sedimentary Structures and Their Hydrodynamic Interpretation*, 133–48. Special publication of the SEPM Society for Sedimentary Geology, 1965. https://doi.org/10.2110/pec.65.08.0133.

Colten, Craig. "Basin Street Blues: Drainage and Environmental Equity in New Orleans, 1890–1930." *Journal of Historical Geography* 28, no. 2 (April 2002): 237–57. https://doi.org/10.1006/jhge.2001.0400.

————. "The Place for Humans in Louisiana Coastal Restoration." *Labor e Engenho* 9, no. 4 (December 2015): 6–18.

Colten, Craig, and Scott Hammering. "Social Impact Assessment Methodology for Diversions and Other Louisiana Coastal Master Plan Restoration and Protection Projects." Produced for and funded by the Coastal Protection and Restoration Authority of Louisiana. Water Institute of the Gulf, February 2014. https://thewaterinstitute.org/assets/docs/reports/4_22_2014_Social-Impact-Assessment-Methodology-for-Diversions-and-other-Louisiana-Coastal-Master-Plan-Projects.pdf.

Condrey, Richard E., Paul E. Hoffman, and Elain D. Evers. "The Last Naturally Active Delta Complexes of the Mississippi River (LNDM): Discovery and Implications." In *Perspectives on the Restoration of the Mississippi Delta: The Once and Future Delta*, edited by John W. Day Jr. et al., 123–40. Dordrecht: Springer, 2014.

Couvillion, Brady R., Holly Beck, Donald Schoolmaster, and Michelle Fischer. "Land Area Change in Coastal Louisiana (1932 to 2016)." Pamphlet to accompany Scientific Investigations Map 3381, U.S. Geological Survey, 2017. Accessed May 17, 2023. https://doi.org/10.3133/sim3381.

Cowardin, Lewis M., Lynn Greenwalt, et al. "Classification of Wetlands and Deepwater Habitats of the United States." FWS/OBS-79/31. US Department of the Interior, Fish and Wildlife Service, Washington, DC, December 1979; repr. 1992. Accessed May 16, 2023. https://www.fws.gov/wetlands/Documents/Classification-of-Wetlands-and-Deepwater-Habitats-of-the-United-States.pdf.

Curole, Wendell. In conversation with the author. Golden Meadow, LA, 2016.

Dahl, Kristina. "What More Gulf of Mexico Oil and Gas Leasing Means for Achieving U.S. Climate Targets." Oversight Hearing before the Subcommittee on Energy and Mineral Resources of the Committee on Natural Resources, U.S. House of Representatives, 117th Cong., 2nd sess. (January 20, 2022) (testimony of Kristina Dahl, Senior Climate Scientist, Union of Concerned Scientists). Accessed May 16, 2023. https://www.congress.gov/event/117th-congress/house-event/LC67912/text?s=1&r=13.

Darensbourg, Jeffery. Speaking at Tulane Gulf South Indigenous Studies Symposium, organized by the New Orleans Center for Gulf South at Tulane University, March 18, 2022.

Dattel, E. R. "Cotton in a Global Economy: Mississippi (1800–1860)." *Mississippi History Now* (October 2006). Accessed May 16. https://www.mshistorynow.mdah.ms.gov/issue /cotton-in-a-global-economy-mississippi-1800-1860.

Davies, T. "Slow Violence and Toxic Geographies: 'Out of Sight' to Whom?" *Environment and Planning C: Politics and Space* 40, no. 2 (2019): 1–19. https://doi .org/10.1177/2399654419841063.

Davis, Donald W. "Historical Perspective on Crevasses, Levees, and the Mississippi River." In *Transforming New Orleans and Its Environs*, edited by Craig Colten, 84–106. Pittsburgh: University of Pittsburgh Press, 2001. https://doi.org/10.2307/j.ctt7zw9kz.12.

———. "Louisiana Canals and Their Influence on Wetland Development." PhD diss., Louisiana State University and Agricultural and Mechanical College, 1973. Accessed May 17, 2023. https://biotech.law.lsu.edu/blog/Don-Davis-Dissertation-on-South-Louisiana-Canals.pdf.

Davis, Heather. *Plastic Matter*. Durham, NC: Duke University Press. 2022.

Davis, Mark. In conversation with the author. New Orleans, September 24, 2018.

Dawdy, Shannon, et al. "Archaeological Investigations at St. Anthony's Garden (16OR443), New Orleans, Louisiana: Volume II: 2009 Fieldwork Results, Faunal Report, Artifact Analyses and Final Site Interpretations." University of Chicago Department of Anthropology, March 2014.

Day, John W., Jr., et al., eds. *Perspectives on the Restoration of the Mississippi Delta: The Once and Future Delta*. Dordrecht: Springer, 2014.

Day, J. W., H. C. Clark, C. Chang, R. Hunter, and C. R. Norman. "Life Cycle of Oil and Gas Fields in the Mississippi River Delta: A Review." *Water* (Switzerland) (2020): 12.

De Bueiw, Edward. "Interview with Edward De Bueiw." WPA Ex-Slave Narrative Project, June 10, 1940. Via Forensic Architecture, "Environmental Racism in Death Alley: Phase 1 Investigative Report." 2021. https://content.forensic-architecture.org/wp-content /uploads/2021/07/Environmental-Racism-in-Death-Alley-Louisiana_Phase-1-Report _Final_2021.07.04.pdf.

DeJong, Allison, Soupy Dalyander, Jessi Parfait, Erin Kiskaddon, Alyssa Dausman, Colleen McHugh, Shubhra Misra, and Scott Hemmerling. "Consequence Analysis of the Draft Portfolio of Climate Strategies and Actions: In Support of the Climate Initiatives Task Force Development of a Louisiana Climate Action Plan." Water Institute of the Gulf, January 21, 2022. Accessed May 18, 2023. https://thewaterinstitute.org/assets/docs/reports /Consequence-Analysis-of-the-Draft-Portfolio-of-Climate-Strategies-and-Actions.pdf.

Diouf, Sylviane A. *Slavery's Exiles: The Story of the American Maroons*. New York: New York University Press, 2014.

Downey, L., and B. Hawkins. "Race, Income, and Environmental Inequality in the United States." *Sociological Perspectives* 51, no. 4 (2008): 759–81. https://doi.org/10.1525/sop .2008.51.4.759.

Du Bois, W. E. B. "The Conservation of Races." In *The Problem of the Color Line at the Turn of the Twentieth Century: The Essential Early Essays*, edited by Nahum Dimitri Chandler. New York: Fordham University Press, 2014. https://doi.org/10.5422/fordham /9780823254545.003.0003.

———. "The Freedman's Bureau." In *The Problem of the Color Line at the Turn of the Twentieth Century: The Essential Early Essays*, edited by Nahum Dimitri Chandler. New York: Fordham University Press, 2014.

Dye, D. "Death March of Hernando de Soto." *Archaeology* 42, no. 3 (May–June 1989): 26–29, 31.

Eggler, Bruce. "FEMA Archeologists Find American Indian Pottery, Other Items by Bayou St. John." *New Orleans Advocate Times-Picayune*, February 21, 2013. Accessed May 16, 2023. https://www.nola.com/news/politics/fema-archeologists-find-american -indian-pottery-other-items-by-bayou-st-john/article_538a51a4-04ed-5ffb-a02d -c526e698ff39.html.

Elie, Eric Lois. "Why We Came Home." *Bitter Southerner*, August 2015. Accessed May 17, 2023. https://bittersoutherner.com/katrina-ten-years-later/why-we-came-home#.ZGTdO -zMI-Q.

Ellis, E. *Great Power of Small Nations*. Philadelphia: University of Pennsylvania Press, 2023.

Ellis, John. *Yellow Fever and Public Health in the New South*. Lexington: University Press of Kentucky, 1992. http://www.jstor.org/stable/j.ctt130hnmm.

Ellwood, Brooks B., Sophie Warny, Rebecca A. Hackworth, Suzanne H. Ellwood, Jonathan H. Tomkin, Samuel J. Bentley, Dewitt H. Braud, and Geoffrey C. Clayton. "The LSU Campus Mounds, with Construction Beginning at ~11,000 BP, Are the Oldest Known Extant Man-Made Structures in the Americas." *American Journal of Science* 322, no. 6 (June 2022): 795–827. https://doi.org/10.2475/06.2022.02.

Environmental Defense Fund. "Groups Pleased as Key Sediment Diversions Advance, Coastal Restoration Funds Protected: CPRA Board Moves Forward on Two Diversion Projects, Proposes Using GOMESA Funds for Highway Elevation." Press release, October 25, 2015. Accessed May 17, 2023. https://www.edf.org/media/groups-pleased -key-sediment-diversions-advance-coastal-restoration-funds-protected.

———. "Mapping Orphan Wells in the State." Accessed May 12, 2023. https://www.edf.org /sites/default/files/2021-10/Orphan%20Well%20FactSheet%20LA.pdf.

Escobar, Arturo. "Construction Nature: Elements for a Post-Structuralist Political Ecology." *Futures* 28, no. 4 (May 1996): 328. https://doi.org/10.1016/0016-3287(96)00011-0.

Evans, Beau. "Forty-Six Tons of Mardi Gras Beads Found in Clogged Catch Basins." *New Orleans Advocate Times-Picayune*, January 25, 2018. Accessed May 17, 2023. https://www .nola.com/news/politics/article_37e0ff53-894c-5aed-b4c3-129852582269.html.

Evans, Freddi Williams. *Congo Square: African Roots in New Orleans*. Lafayette: University of Louisiana at Lafayette Press, 2011.

Faber, Lo. "Slave Insurrections in Louisiana." 64 Parishes, October 13, 2011. Accessed May 17, 2023. https://64parishes.org/entry/slave-insurrections-in-louisiana.

"Fazendeville." Jean Lafitte National Historical Park and Preserve. Accessed May 16, 2023. https://www.nps.gov/jela/learn/historyculture/places-fazendeville.htm.

Fazio, Marie. "'McDonough Three' Integrated New Orleans Schools. Now, They're Part of the Civil Rights Trail." *New Orleans Advocate Times-Picayune*, February 1, 2022. Accessed May 17, 2023. https://www.nola.com/news/education/article_d507518c-839c -11ec-8d00-ff3a062a1d6a.html.

Federal Writers' Project. *New Orleans City Guide*. Written and compiled by the Federal Writers' Project of the Works Progress Administration for the City of New Orleans. Boston: Houghton Mifflin, 1938.

Festa, David. "In a Race against Time, Officials Collaborate to Speed Up Coastal Restoration. Here's How." *Growing Returns* (blog), Environmental Defense Fund, April 23,

2018. Accessed May 17, 2023. http://blogs.edf.org/growingreturns/2018/04/19/officials
-collaborate-to-speed-up-coastal-restoration.

Fiehrer, Thomas Marc. "The Baron de Carondelet as Agent of Bourbon Reform: A Study of Spanish Colonial Administration in the Years of the French Revolution." PhD diss., Tulane University, 1977.

Fields, Gary. "'This Is *Our* Land': Collective Violence, Property Law, and Imagining the Geography of Palestine." *Journal of Cultural Geography* 29, no. 3 (October 2012): 267–91. https://doi.org/10.1080/08873631.2012.726430.

Fisk, Harold. "Geological Investigation of the Alluvial Valley of the Lower Mississippi River." Conducted for the Mississippi River Commission, US Army Corps of Engineers, Vicksburg, MS, December 1, 1944, https://ngmdb.usgs.gov/Prodesc/proddesc_70640 .htm.

"500+ Organizations Call on US and Canadian Leaders to Reject Carbon Capture and Storage as a False Solution to Climate Crisis." Waterkeeper Alliance, July 19, 2021. Accessed May 16, 2023. https://waterkeeper.org/news/500-organizations-call-on-us-and-canadian -leaders-to-reject-carbon-capture-and-storage-as-a-false-solution-to-climate-crisis/.

Flatt, Victor Byers. "Gasping for Breath: The Administrative Flaws of Federal Hazardous Air Pollution Regulation and What We Can Learn from the States." *Ecology Law Quarterly* 34 (2007): 107–73.

Fleischman, L., and M. Franklin. *Fumes across the Fence-Line: The Health Impacts of Air Pollution from Oil and Gas Facilities on African American Communities.* Baltimore, MD: NAACP, Clean Air Task Force, 2017.

Follet, Richard. *The Sugar Masters: Planters and Slaves in Louisiana's Cane World, 1820–1860.* Baton Rouge: LSU Press, 2005.

Forbes, Jack D. "The Classification of Native Americans as Mulattoes." In *Africans and Native Americans.* Urbana: University of Illinois Press, 1993. Accessed May 17, 2023. https://www.uib.no/sites/w3.uib.no/files/attachments/forbes_2-africans_and_native _americans.pdf.

———. "Mulattoes and People of Color in Anglo-North America: Implications for Black-Indian Relations." *Journal of Ethnic Studies* 12, no. 2 (Summer 1984): 17–61.

Forensic Architecture. "Environmental Racism in Death Alley: Phase 1 Investigative Report." 2021. https://content.forensic-architecture.org/wp-content/uploads/2021/07 /Environmental-Racism-in-Death-Alley-Louisiana_Phase-1-Report_Final_2021.07.04 .pdf.

Foucault, Michel. *Security, Territory, Population: Lectures at the Collège de France, 1977–78.* Edited by Michel Senellart. Translated by Graham Burchell. Basingstoke, UK: Palgrave Macmillan, 2007.

———. "Two Lectures" and "Truth and Power." In *Power/Knowledge—Selected Interviews and Other Writings, 1972–1977,* edited by Colin Gordon. 1976; New York: Pantheon, New York: 1980.

Freudenburg, William, et al. "Disproportionality and Disaster: Hurricane Katrina and the Mississippi River-Gulf Outlet." *Social Science Quarterly* 90, no. 3 (September 2009): 497–515.

Gagliano, Sherwood. "Canals, Dredging, and Land Reclamation in the Louisiana Coastal Zone." In *Hydrologic and Geologic Studies of Coastal Louisiana, Report 14.* Baton Rouge:

LSU Center for Wetland Resources, October 1973. https://www.govinfo.gov/content/pkg/CZIC-gc57-2-l667-no-14/html/CZIC-gc57-2-l667-no-14.htm.

Gagliano, Sherwood, Phillip Light, and Ronald Becker. "Controlled Diversions in the Mississippi Delta System: An Approach to Environmental Management." In *Hydrologic and Geologic Studies of Coastal Louisiana, Report* 8. Baton Rouge: LSU Center for Wetland Resources Coastal Resources Unit, 1973.

Gagliano, Sherwood, Klaus Meyer-Arendt, and Karen Wicker. "Land Loss in Mississippi River Deltaic Plain." *AAPG Bulletin* 65, no. 9 (September 1981): 1684.

Ghosh, Amitav. *The Great Derangement: Climate Change and the Unthinkable*. Chicago: University of Chicago Press, 2017.

Gibbons v. Ogden, 22 U.S. 1 (1824). Accessed May 12, 2023. https://supreme.justia.com/cases/federal/us/22/1/.

Giblett, Rodney James. *Postmodern Wetlands: Culture, History, Ecology*. Edinburgh: Edinburgh University Press, 1996. http://www.jstor.org/stable/10.3366/j.ctvxcrp94.

Giosan, Liviu, and Angelina Freeman. "How Deltas Work: A Brief Look at the Mississippi River Delta in a Global Context." In *Perspectives on the Restoration of the Mississippi Delta: The Once and Future Delta*, edited by John W. Day Jr. et al. Dordrecht: Springer, 2014.

Gobert, Lenora, Anne Rolfes, and Justin Kray. "Genealogy Zoom Call. July 7, 2021." YouTube video, 1:21:45. Posted by Justin Kray, July 8, 2021. Accessed May 12, 2023. https://www.youtube.com/watch?v=XomIx88YoNM.

Goolsby, Donald. "Mississippi Basin Nitrogen Flux Believed to Cause Gulf Hypoxia." *Eos: Transactions* 81, no. 29 (July 18, 2000): 321, 326–27.

Grabar, Henry. "Hell Is High Water: When Will the Mississippi River Come for New Orleans?" *Slate*, June 18, 2019. Accessed May 16, 2023. https://slate.com/news-and-politics/2019/06/mississippi-river-new-orleans-flood-control-arkansas-tennessee-missouri-illinois.html.

Grant, Richard. "Deep in the Swamps, Archaeologists Are Finding How Fugitive Slaves Kept Their Freedom." *Smithsonian Magazine*, September 2016. Accessed May 16, 2023. https://www.smithsonianmag.com/history/deep-swamps-archaeologists-fugitive-slaves-kept-freedom-180960122/.

Gudmestad, Robert. "Steamboats and the Removal of the Red River Raft." *Louisiana History: Journal of the Louisiana Historical Association* 52, no. 4 (2011): 389–416.

Hall, A. R. "Public Slaves and State Engineers: Modern Statecraft on Louisiana's Waterways, 1833–1861." *Journal of Southern History* 85, no. 3 (2019): 531–76.

Hall, Gwendolyn Midlo. *Africans in Colonial in Louisiana: The Development of Afro-Creole Culture in the Eighteenth Century*. Baton Rouge: LSU Press, 1992.

Hammer, David. "What Is Causing Louisiana's Insurance Crisis, and What Can Fix It?" 4WWL.com, February 2, 2023. Accessed May 16, 2023. https://www.wwltv.com/article/news/investigations/david-hammer/louisianas-insurance-crisis-what-can-fix-it/289-a9fe2f3c-8701-4f75-959f-6ec7b5e7f380.

Hannah-Jones, Nikole, Khalil Gibran Muhammad, et al. "The 1619 Project." *New York Times Magazine*, August 14, 2019. Accessed May 16, 2023. https://www.nytimes.com/interactive/2019/08/14/magazine/sugar-slave-trade-slavery.html.

Haraway, Donna. *Staying with the Trouble: Making Kin in the Chthulucene*. Durham, NC: Duke University Press, 2016.

Harrison, R. *Levee Districts and Levee Building in Mississippi: A Study of State and Local Efforts to Control Mississippi River Floods.* Stoneville, MS: Delta Council, 1951.

Haselle, Della. "Voluntary Relocation, Construction Limits among the Options to Deal with Rising Water along Louisiana Coast." *The Lens*, February 15, 2018. Accessed May 17, 2023. https://thelensnola.org/2018/02/15/voluntary-relocation-construction-limits-among -the-options-to-deal-with-rising-water-along-louisiana-coast/.

Healy, Jack. "Black, Native American and Fighting for Recognition in Indian Country." *New York Times*, September 8, 2020. Accessed May 16, 2023. https://www.nytimes .com/2020/09/08/us/enslaved-people-native-americans-oklahoma.html.

Hebert, Josef. "Katrina Spurs New Debate on Energy, Fuel Economy, Offshore Drilling." Associated Press, September 12, 2005. Accessed May 13, 2023. https://www.enn.com /articles/2587-katrina-spurs-new-debate-on-energy,-fuel-economy,-offshore-drilling.

Helman, Christopher. "Open the Spigots." *Forbes*, October 3, 2005. Accessed May 13, 2023. https://www.forbes.com/free_forbes/2005/1003/049.html.

Hess, Amanda. "Apocalypse When? Global Warming's Endless Scroll." *New York Times*, February 3, 2022. Accessed May 13, 2023. https://www.nytimes.com/2022/02/03/arts /climate-change-doomsday-culture.html.

Hilburn, Greg. "How Louisiana Became the Carbon Capture Capital of the South with $6 Billion in New Projects." *Lafayette Daily Advertiser*, April 17, 2022. Accessed May 13, 2023. https://www.theadvertiser.com/story/news/2022/04/18/how-louisiana-became-carbon -capture-capital-south-john-bel-edwards-bill-cassidy-air-products-cleco/7330762001/.

"History of Alexandria, the Early Years." Alexandria-Louisiana.com. Accessed April 24, 2023. https://www.alexandria-louisiana.com/index.htm.

Hite, Elizabeth Ross. "Interview with Elizabeth Ross Hite." WPA Ex-Slave Narrative Project, ca. 1940. Via Forensic Architecture, "Environmental Racism in Death Alley: Phase 1 Investigative Report." 2021. https://content.forensic-architecture.org/wp -content/uploads/2021/07/Environmental-Racism-in-Death-Alley-Louisiana_Phase -1-Report_Final_2021.07.04.pdf.

Hollandsworth, James G. "The Burning of Alexandria." 64 Parishes. Accessed May 13, 2023. https://64parishes.org/burning-alexandria.

Hosbey, Justin, and J. T. Roane. "A Totally Different Form of Living: On the Legacies of Displacement and Marronage as Black Ecologies." *Southern Cultures* 27, no. 1 (Spring 2021): 68–73. https://doi.org/10.1353/SCU.2021.0009.

Houck, Oliver A. In conversation with the author. 2017.

———. "The Reckoning : Oil and Gas Development in the Louisiana Coastal Zone." *Tulane Environmental Law* 28, no. 2. (2015): 185–296.

———. "Shintech: Environmental Justice at Ground Zero." *Georgetown Environmental Law Review* 31, no. 3 (Spring 2019): 455–507.

How, Louis. *James B. Eads.* Boston: Houghton, Mifflin and Co., 1900; Project Gutenberg, 2008.

Howarth, Robert, and Mark Jacobson. "How Green Is Blue Hydrogen?" *Energy Science and Engineering* 9, no. 10 (2021): 1676–87. https://doi.org/10.1002/ese3.956.

Hughes, Thomas P. *Human-Built World: How to Think about Technology and Culture.* Chicago: University of Chicago Press, 2005.

Hunter, Rachael, et al. "Using Natural Wetlands for Municipal Effluent Assimilation: A Half-Century of Experience for the Mississippi River Delta and Surrounding Environs."

In *Multifunctional Wetlands: Pollution Abatement and Other Ecological Services from Natural and Constructed Wetlands*, edited by Nidhi Nagabhatla and Christopher D. Metcalfe, 15–81. Cham, Switzerland: Springer, 2018.

"Hydrology Report for Project Minerva." Accessed April 24, 2023. https://www.documentcloud.org/documents/21070086-hydrology-report-for-project-minerva.

"Hyperobjects, Hyposubjects and Solidarity in the Anthropocene: Anthropocenes Interview with Timothy Morton and Dominic Boyer." *Anthropocenes—Human, Inhuman, Posthuman* 1, no. 1 (2020). https://doi.org/10.16997/ahip.5.

"Hypoxia 101." Mississippi River/Gulf of Mexico Hypoxia Task Force, U.S. Environmental Protection Agency. Accessed April 6, 2017. https://www.epa.gov/ms-htf/hypoxia-101.

Independent Levee Investigation Team. "Geology of the New Orleans Region." Chapter 3 in *Investigation of the Performance of the New Orleans Flood Protection Systems in Hurricane Katrina on August 29, 2005*. National Science Foundation and CITRIS at the University of California at Berkeley, July 31, 2006. https://usace.contentdm.oclc.org/digital/collection/p266001coll1/id/2935/.

Inside Energy Extra. "Domenici Plans New Drive on OCS." September 6, 2005, 1.

Intergovernmental Panel on Climate Change. "Sixth Assessment Report: Working Group 1; The Physical Science Basis." Accessed June 4, 2022. https://www.ipcc.ch/report/ar6/wg1/.

Isaacson, Walter. "The Greatest Education Lab: How Katrina Opened the Way for an Influx of School Reformers." *Time*, September 6, 2007. Accessed May 13, 2023. http://content.time.com/time/subscriber/article/0,33009,1659767,00.html.

Jackson, Jeremy, and Steve Chapple. *Breakpoint: Reckoning with America's Environmental Crises*. New Haven, CT: Yale University Press, 2018.

Jackson, Steven J. "Building the Virtual River: Numbers, Models, and the Politics of Water in California." PhD diss., University of California San Diego, 2005.

Jacobs, K. "Monitoring and Enforcement of Air Quality: Department of Environmental Quality." *LLA Reports Podcast*, 2021. Accessed May 13, 2023. https://www.lla.la.gov/reports/podcasts.

Jessee, Nathan. "Reshaping Louisiana's Coastal Frontier: Managed Retreat as Colonial Decontextualization." *Journal of Political Ecology* 29, no. 1 (2022): 277–301.

Johnson, Cedric, ed. *The Neoliberal Deluge: Hurricane Katrina, Late Capitalism, and the Remaking of New Orleans*. Minneapolis: University of Minnesota Press, 2011.

Johnston, Bennett J. *Old River Control Structure, Louisiana: Hearing before a Subcommittee of the Committee on Appropriations, Special Hearing: Corps of Engineers—Civil Nondepartmental Witnesses*, 96th Cong., 2nd sess. (December 22, 1980) (statement of J. Bennett Johnston, US Senator of Louisiana).

Johnston, James B., Donald Cahoon, and Megan La Peyre. "Technical Summary for 'Outer Continental Shelf (OCS)-Related Pipelines and Navigation Canals in the Western and Central Gulf of Mexico: Relative Impacts on Wetland Habitats and Effectiveness of Mitigation' (OCS Study MMS 2009-048)." Access no. 14961. US Department of the Interior, Minerals Management Service, Gulf of Mexico OCS Region, New Orleans, September 2009.

Joselow, Maxine. "Top Companies Are Undermining Their Climate Pledges with Political Donations, Report Says." *Washington Post*, February 23, 2022. Accessed May 13, 2023.

https://www.washingtonpost.com/politics/2022/02/23/top-companies-are-undermining-their-climate-pledges-with-political-donations-report-says/.

Kang, Mary. "Workshop: Analyzing the Challenges of Improperly Abandoned and Orphaned Wells." American Association for the Advancement of Science, October 21, 2020. Accessed May 13, 2023. https://www.aaas.org/events/workshop-analyzing-challenges-improperly-abandoned-and-orphaned-wells.

Kaplan-Levenson, Laine. "New Orleans: 300 // Bulchancha: 3000." *TriPod: New Orleans at 300*, WWNO (New Orleans Public Radio), December 20, 2018. Accessed May 16, 2023. https://www.wwno.org/post/new-orleans-300-bulbancha-3000.

Kardas-Nelson, Molly. "The Petrochemical Industry Is Killing Another Black Community in 'Cancer Alley.'" *The Nation*, August 26, 2019.

Katz, William Loren. *Black Indians: A Hidden Heritage.* 1st ed. New York: Atheneum, 1986.

Kearney, Michael, Alex Rider, and R. Eugene Turner. "Freshwater River Diversions for Marsh Restoration in Louisiana: Twenty-Six Years of Changing Vegetative Cover and Marsh Area." *Geophysical Research Letters* 38, no. 16 (August 2011): L16405. https://doi.org/10.1029/2011GL047847.

Keddy, P. A., et al. "The Wetlands of Lakes Pontchartrain and Maurepas: Past, Present and Future." *Environmental Reviews* 15 (December 2007): 53–81. https://doi.org/10.1139/a06-008.

Kelman, Ari. "Boundary Issues: Clarifying New Orleans's Murky Edges." *Journal of American History* 94, no. 3 (December 2007): 695–703.

Klein, Christine, and Sandra B. Zellmer. *Mississippi River Tragedies: A Century of Unnatural Disaster.* New York: New York University Press, 2016.

Klein, Naomi. *The Shock Doctrine: The Rise of Disaster Capitalism.* New York: Picador, 2008.

Kniffen, Fred, Gregory Hiram, and George Stokes. *The Historic Indian Tribes of Louisiana: From 1542 to the Present.* Baton Rouge: LSU Press, 1987.

Knowles, Lindsay. "Four More Dead Dolphins Wash Ashore between Long Beach and Ocean Springs." *Fox 8 Live*, May 28, 2019. Accessed May 16, 2023. https://www.wlox.com/2019/05/28/three-more-dead-dolphins-wash-ashore-between-long-beach-ocean-springs/.

Kolker, Alex, Mead A. Allison, and Sultan Hameed. "An Evaluation of Subsidence Rates and Sea-Level Variability in the Northern Gulf of Mexico." November 11, 2011. *Geophysical Research Letters*, vol. 38, L21404. DOI: https://doi.org/10.1029/2011GL049458.

Lamb, William, Giacomo Mattioli, Sara Levi, James Roberts, Steve Capstick, Felix Creutzig, Corinne Le Quéré, et al. "Discourses of Climate Delay." *Global Sustainability* 3 (2020): E17. https://doi.org/10.1017/sus.2020.13.

Latour, Bruno, and Catherine Porter. *We Have Never Been Modern.* Cambridge, MA: Harvard University Press, 1993.

Laurent, Lubin. "A History of St. John the Baptist Parish." *Louisiana Historical Quarterly* 7 (1924).

Lavigne, Sharon. "Formosa Plastic Toxic Tour." Zoom Call. June 30, 2021.

Laville, Sandra. "BHP Reveals Five Mine Dams at 'Extreme' Risk of Causing Damage and Loss of Life." *The Guardian*, June 7, 2019. Accessed May 3, 2023. https://www.theguardian.com/business/2019/jun/07/bhp-reveals-five-mining-dams-at-extreme-risk-of-causing-damage-and-loss-of-life.

Layton, Edwin. "Mirror-Twins: The Communities of Science and Technology in 19th-Century America." *Technology and Culture* 12, no. 4 (October 1971): 562–80.

Leavitt, Mel. *A Short History of New Orleans*. San Francisco: Lexikos, 1982.

Leber, Rebecca. "Your Plastic Addiction Is Bankrolling Big Oil." *Mother Jones*, March–April 2020, 1–3. Accessed May 1, 2023. https://www.motherjones.com/environment/2020/03 /your-plastic-addiction-is-bankrolling-big-oil/.

LeBlanc, Rufus, Sr. "Harold Norman Fisk as a Consultant to the Mississippi River Commission, 1948–1964—an Eye-Witness Account." *Engineering Geology* 45, nos. 1–4 (December 1996): 15–36.

Legendre, Jordon. "Coastal Plan Could Help Diversify Economy." *Houma Today*, February 3, 2017. Accessed May 13, 2023. https://www.houmatoday.com/news/20170202/coastal -plan-could-help-diversify-economy.

Leithart, Peter. "We Have Never Been Modern." *Theopolis*, October 23, 2013. Accessed May 16, 2023. https://theopolisinstitute.com/we-have-never-been-modern/.

Le Miere, Jason. "Donald Trump Says 'Our Ancestors Tamed a Continent' and 'We Are Not Going to Apologize for America.'" *Newsweek*, May 25, 2018.

Lerner, Steve. *Diamond: A Struggle for Environmental Justice in Louisiana's Chemical Corridor*. Cambridge, MA: MIT Press, 2005.

———. *Sacrifice Zones*. Cambridge, MA: MIT Press, 2010. https://doi.org/10.7551 /mitpress/8157.001.0001.

Levantesi, Stella. "Climate Deniers and the Language of Climate Obstruction." *DeSmog*, June 16, 2022. Accessed May 13, 2023. https://www.desmog.com/2022/06/16/climate -deniers-fossil-fuel-language-obstruction/.

Lewis, Pierce. *New Orleans: The Making of an Urban Landscape*. Charlottesville: University of Virginia Press, 2002.

Lin, Rong-Gong. "Louisianans Rally against Oil Drilling Moratorium." *Los Angeles Times*, July 22, 2010. Accessed May 13, 2023. https://www.latimes.com/archives/la-xpm-2010 -jul-22-la-na-0722-oil-spill-rally-20100722-story.html.

Lindner, Taylor. "Sugar Granulation on the Boré Plantation." Edited by Kalie A. Dutra and Kathryn O'Dwyer. *New Orleans Historical*. Accessed May 30, 2022. https:// neworleanshistorical.org/items/show/1655.

Lippincott, Isaac. "A History of River Improvement." *Journal of Political Economy* 22, no. 7 (July 1914): 630–60.

Lopez, John. "The Multiple Lines of Defense Strategy to Sustain Coastal Louisiana." In "Geologic and Environmental Dynamics of the Pontchartrain Basin," special issue, *Journal of Coastal Research*, no. 54 (Fall 2009): 186–97. http://www.jstor.org/stable/25737479.

Louisiana Bucket Brigade. "Down by the River." Accessed April 24, 2023. https:// labucketbrigade.org/our-work/down-by-the-river/.

"Louisiana Climate Action Plan: Climate Initiatives Task Force Recommendations to the Governor." February 2022. Accessed May 16, 2023. https://gov.louisiana.gov/assets/docs /CCI-Task-force/CAP/Climate_Action_Plan_FINAL_3.pdf.

Louisiana Coastal Wetlands Conservation and Restoration Task Force. "Louisiana Coastal Wetlands Restoration Plan: Main Report and Environmental Impact Statement." Baton Rouge, November 1993. Accessed May 16. https://la.dwh.com/wp-content /uploads/2018/02/8.2.4.6.1.3.1_LCWCRTF.1993_LaCoastal_Wetlands_RestorationPlan _Main_ReportEIS.pdf.

Louisiana Coastal Wetlands Conservation and Restoration Task Force and the Wetlands Conservation and Restoration Authority. "Coast 2050: Toward a Sustainable Coastal Louisiana, An Executive Summary." Baton Rouge: Louisiana Department of Natural Resources, 1998. Accessed May 16, 2023. https://www.webharvest.gov /peth04/20041015003758/http://lacoast.gov/Programs/2050/MainReport/report1.pdf.

Louisiana Office of the Governor. "Gov. Edwards Requests Federal Disaster Declaration for Flooded Fisheries." June 17, 2019. Accessed October 30, 2020. http://gov.louisiana.gov /index.cfm/newsroom/detail/1995.

Louisiana Recovery Authority. "Progress Report." December 2007. Accessed May 16, 2023. http://lra.louisiana.gov/assets/docs/searchable/Quarterly%20Reports/December2007 QtReport.pdf.

Louisiana State University. "A Brief History of LSU Geology and Geophysics." Accessed June 14, 2019. https://www.lsu.edu/science/geology/about_lsu_geology/program_history /index.php.

Lyon, Edwin Austin, II. "New Deal Archaeology in the Southeast: WPA, TVA, NPS, 1934–1942." PhD diss., Louisiana State University and Agricultural and Mechanical College, 1982. Accessed May 13, 2023. https://digitalcommons.lsu.edu/gradschool_disstheses/3728.

———. *A New Deal for Southeastern Archaeology*. Tuscaloosa: University of Alabama Press, 1996.

Lyons, Kristina Marie. "Decomposition as Life Politics: Soils, Selva, and Small Farmers under the Gun of the U.S.-Colombian War on Drugs." *Cultural Anthropology* 31, no. 1 (February 2016): 56–81. https://doi.org/10.14506/ca31.1.04.

Magelssen, Scott. *Simming: Participatory Performance and the Making of Meaning*. Ann Arbor: University of Michigan Press, 2014. https://www.jstor.org/stable/10.3998/mpub .4969005?turn_away=true.

Maldonado, Julie K. "A Multiple Knowledge Approach for Adaptation to Environmental Change: Lessons Learned from Coastal Louisiana's Tribal Communities." *Journal of Political Ecology* 21, no. 1 (2014): 61–82. https://doi.org/10.2458/v21i1.21125.

———. *Seeking Justice in an Energy Sacrifice Zone: Standing on Vanishing Land in Coastal Louisiana*. New York: Taylor and Francis, 2018.

Mallach, Alan. "Where Will People Live? New Orleans' Growing Rental Housing Challenge." Prepared for the New Orleans Redevelopment Authority. Center for Community Progress, June 2016. Accessed May 13, 2023. https://www.communityprogress.net /filebin/NORA-Rental-Housing-Report-final_6_20_16.pdf.

Manders, Damon. "The US Army Corps of Engineers and the Mississippi River Cutoff Plan." In *Engineering Earth: The Impacts of Megaengineering Projects*, edited by Stanley D. Brunn, 1451–63. Dordrecht: Springer, 2011.

Marsalis, Wynton. "Saving America's Soul Kitchen." *Time*, September 12, 2005. Accessed May 13, 2023. https://wyntonmarsalis.org/news/entry/wyntons-article-on-time-magazine -today.

Martin, C. W., L. O. Hollis, and R. E. Turner. "Effects of Oil-Contaminated Sediments on Submerged Vegetation: An Experimental Assessment of *Ruppia maritima*." *PLoS ONE* 10, no. 10 (2015): e0138797. https://doi.org/10.1371/journal.pone.0138797.

Martin, Jerry. "Contracting for Louisiana Coastal Restoration." *1012 Industry Report*, September 22, 2015. Accessed May 13, 2023. https://www.1012industryreport.com /environmental/contracting-louisiana-coastal-restoration/.

Marx, Daniela, and Shannon Lee Dawdy. "La Village des Chapitoulas." Paper Monuments Project #014, New Orleans Historical. Accessed April 24, 2019. https://neworleanshistorical.org/items/show/1404.

Mathewson, Kent, and Vincent Shoemaker. "Louisiana State University Geography at Seventy-Five: 'Berkeley on the Bayou' and Beyond." In *The Role of the South in the Making of American Geography: Centennial of the AAG*, edited by James O. Wheeler and Stanley D. Brunn, 245–67. Columbia, MD: Bellwether, 2004.

Mathis, Ruth, and Terry Weik. "Not Just Black and White: African Americans Reclaiming the Indigenous Past." In *Indigenous Archaeologies: Decolonizing Theory and Practice*, edited by Caire Smith and H. Martin Wobst. New York: Routledge, 2005.

McConnaughey, Janet. "History in Bloom: Louisiana Irises, African Lilies May Mark Razed Black Community." Associated Press, April 3, 2022. Accessed May 16, 2023. https://apnews.com/article/travel-new-orleans-plants-war-casualties-environment-8a48af200a5cd2e0028d91e9b60c35e7.

McDowell, Robin. "Sacred Ground: Unearthing Buried History at the Bonnet Carré Spillway." *AntiGravity Magazine*, May 2019. Accessed April 24, 2023. https://antigravity magazine.com/feature/sacred-ground-unearthing-buried-history-at-the-bonnet-carre-spillway/.

McKenna, Phil. "The IPCC Understated the Need to Cut Emissions from Methane and Other Short-Lived Climate Pollutants, Climate Experts Say." *Inside Climate News*, August 12, 2021. Accessed April 24, 2023. https://insideclimatenews.org/news/12082021/ipcc-report-methane-super-pollutants.

McLindon, Chris. "Oil and Gas Geologist: Suing Energy Industry Won't Help Preserve New Orleans." *Advocate* (Baton Rouge), April 16, 2019. Accessed April 24, 2023. https://www.theadvocate.com/baton_rouge/opinion/article_75060ee0-6059-11e9-9e4f-33d83a94e673.html.

McMichael, R. N. "Plant Location Factors in the Petrochemical Industry in Louisiana." PhD diss., Louisiana State University and Agricultural and Mechanical College, 1961. Accessed May 16, 2023. https://digitalcommons.lsu.edu/gradschool_disstheses/695.

McPhee, John. "Atchafalaya." *New Yorker*, February 23, 1987. Accessed April 24, 2023. https://www.newyorker.com/magazine/1987/02/23/atchafalaya.

Meiners, J. "Polluters Paradise: How Oil Companies Avoided Environmental Accountability after 10.8 Million Gallons Spilled." *ProPublica* and *New Orleans Advocate Times-Picayune*, 2019. Accessed April 24, 2023. https://www.propublica.org/article/polluters-paradise-how-oil-companies-avoided-environmental-accountability-after-10-8-million-gallons-spilled.

Miller, David. *Dark Eden: The Swamp in Nineteenth-Century American Culture*. Cambridge: Cambridge University Press, 2010.

Miller, Toby. *Greenwashing Culture*. London: Routledge, 2018.

"Mississippi River Sediment Delivery System—Bayou Dupont (BA-39)." Louisiana Coastal Wetlands Conservation and Restoration Task Force, April 2016. Accessed May 16, 2023. https://lacoast.gov/reports/gpfs/BA-39.pdf.

"Mississippi River—Ship Marine Traffic Live Tracking AIS Map Density Map. Ships Current Position." Marinevesseltraffic.com, May 30, 2022. Accessed April 24, 2023. https://www.marinevesseltraffic.com/MISSISSIPPI-RIVER/ship-traffic-tracker.

Mitchell, David. "Corps Will Take Closer Look at Formosa Plant's Impact on Environment, Minority Residents in St. James." *The Advocate* (Baton Rouge), August 18, 2021. Accessed May 17, 2023. https://www.theadvocate.com/baton_rouge/news/article_9b6c9af6-0046 -11ec-b795-e7f65cfad736.html.

———. "'Forever Chemicals' Found in Lower Mississippi. More Testing, Regulation Needed?" *The Advocate* (Baton Rouge), January 23, 2023. Accessed May 16, 2023. https://www .theadvocate.com/baton_rouge/news/forever-chemicals-found-in-lower-mississippi /article_7a123d4a-983c-11ed-8ba4-e7d249565afd.html.

———. "Judge Tosses Air Permit for Giant Plastics Complex, Citing Potential Impact on Black Community." *The Advocate* (Baton Rouge), September 24, 2022. Accessed May 17, 2023. https://www.theadvocate.com/baton_rouge/news/judge-tosses-air-permit-for-giant -plastics-complex-citing-potential-impact-on-black-community/article_794b08d6 -3466-11ed-9723-37c7855e4467.html.

Mitchell, Timothy. *Rule of Experts: Egypt, Techno-Politics, Modernity.* Berkeley: University of California Press, 2002.

Mitman, Gregg. "Reflections on the Plantationocene: A Conversation with Donna Haraway and Anna Tsing." *Edge Effects*, June 18, 2019. Accessed May 17, 2023. https://edgeeffects .net/haraway-tsing-plantationocene/.

Moertle, Randy. Telephone interview by the author. New Orleans, August 31, 2018.

Moore-O'Neal, Wendi. Anthropocene River Campus: The Human Delta at Tulane University. New Orleans, November 2019.

Morris, Christopher. *The Big Muddy: An Environmental History of the Mississippi and Its Peoples from Hernando de Soto to Hurricane Katrina.* New York: Oxford University Press, 2012.

———. "Only a River." *Iowa Review* 39, no. 2 (2009): 149–65. https://doi.org/10.17077/0021 -065X.6718.

Morton, Robert, and Julie C. Bernier. "Recent Subsidence-Rate Reductions in the Mississippi Delta and Their Geological Implications." *Journal of Coastal Research* 26, no. 3 (May 2010): 559–69. https://doi.org/10.2112/jcoastres-d-09-00014r1.1.

Morton, Robert, Julie Bernier, and John Barras. "Evidence of Regional Subsidence and Associated Interior Wetland Loss Induced by Hydrocarbon Production, Gulf Coast Region, USA." *Environmental Geology* 50, no. 2 (May 2006): 261–74. https://doi.org/10.1007 /s00254-006-0207-3.

Morton, Robert A., Noreen Buster, and Dennis Krohn. "Subsurface Controls on Historical Subsidence Rates and Associated Wetland Loss in Southcentral Louisiana." *Transactions: Gulf Coast Association of Geological Societies* 52 (2002): 767–78.

Morton, Timothy. *Hyperobjects: Philosophy and Ecology after the End of the World.* Minneapolis: University of Minnesota Press, 2013.

Mukerji, Chandra. *Modernity Reimagined: An Analytic Guide.* New York: Routledge, 2017.

Murphy, Paul. "Maneuvers on the Mississippi River Difficult as Waters Rise." WWL-TV, March 7, 2018. Accessed May 16, 2023. https://www.wwltv.com/article/news/local /maneuvers-on-the-mississippi-river-difficult-as-waters-rise/289-526764668.

Muth, David P. "The Once and Future Delta." In *Perspectives on the Restoration of the Mississippi Delta: The Once and Future Delta,* edited by John W. Day Jr. et al., 45–65. Dordrecht: Springer, 2014.

Nair, U. S., E. Rappin, E. Foshee, W. Smith, R. A. Pielke, R. Mahmood, J. L. Case, C. B. Blankenship, M. Shepherd, J. A. Santanello, and D. Niyogi. "Influence of Land Cover and Soil Moisture Based Brown Ocean Effect on an Extreme Rainfall Event from a Louisiana Gulf Coast Tropical System." *Scientific Reports* 9 (2019). https://doi.org/10.1038/s41598 -019-53031-6.

Natter, Ari. "Ida Leaves Toxic Chemicals, Oil Spills, and Sewerage Swirling in her Wake." Reuters, September 3, 2021. Accessed May 17, 2023. https://gcaptain.com/ida-toxic-oil -spills/.

Neill, C., and R. E. Turner. "Backfilling Canals to Mitigate Wetland Dredging in Louisiana Coastal Marshes." *Environmental Management* 11 (1987): 823–36.

Nevius, Marcus P. "New Histories of Marronage in the Anglo-Atlantic World and Early North America." *History Compass* 18, no. 5 (2020). https://doi.org/10.1111/hic3.12613.

NOAA Research News. "Greenhouse Gas Pollution Trapped 49% More Heat in 2021 Than in 1990, NOAA Finds." NOAA Global Monitoring Laboratory, Annual Greenhouse Gas Index, May 23, 2022. Accessed May 16, 2023. https://gml.noaa.gov/aggi/aggi.html.

O'Donnell, S. "Management Alert: Prompt Action Needed to Inform Residents Living Near Ethylene Oxide Emitting Facilities about Health Concerns and Actions to Address Those Concerns." US Environmental Protection Agency, March 31, 2020. Accessed May 17, 2023. https://www.epa.gov/sites/production/files/2020-03/documents /_epaoig_20200331-20-n-0128_0.pdf.

O'Neill, Karen. *Rivers by Design: State Power and the Origins of U.S. Flood Control.* Durham, NC: Duke University Press, 2006.

Pabis, George. "Delaying the Deluge: The Engineering Debate over Flood Control on the Lower Mississippi River, 1846–1861." *Journal of Southern History* 64, no. 3 (August 1998): 421–54.

Pappas, Stephanie. "Why Did Hurricane Ida Stay So Strong for So Long?" *Live Science*, August 30, 2021. Accessed May 17, 2023. https://www.livescience.com/hurricane-ida -brown-ocean.html.

Parker, Halle. "Blue Hydrogen Plant Touted for Louisiana, but Will It Reduce Carbon Emissions?" *New Orleans Advocate Times-Picayune*, November 3, 2021. Accessed May 16, 2023. https://www.airproducts.com/company/innovation/carbon-capture#/.

Pearcy, Matthew. "A History of the Ransdell-Humphreys Flood Control Act of 1917." *Louisiana History: Journal of the Louisiana Historical Association* 41, no. 2 (Spring 2000): 113–59.

Peltz, Adam. Comments to proposed rule regulating orphan wells. Potpourri, Louisiana Registry, March 20, 2023.

Penland, Shea, et al., cartographers. *Process Classification of Coastal Land Loss between 1932 and 1990 in the Mississippi River Delta Plain, Southeastern Louisiana.* USGS Open-File Report 00-418. U.S. Geological Survey, 2001. http://pubs.usgs.gov/of/2000/of00-418 /of00-418.pdf.

Persaud, P. "Louisiana EarthQuakes." LSU Office of Research and Economic Development, April 2022. Accessed May 17, 2023. https://www.youtube.com/channel /UCDfWUCWXAeVA2N1kYtCtTUg?mc_cid=95184e7d6e&mc_eid=9a7af8035e.

Peterson, R. W. *Giants on the River: A Story of Chemistry and the Industrial Development on the Lower Mississippi River Corridor.* Baton Rouge, LA: Homesite Co., 2000.

Pezzullo, Phaedra C. "Resisting Carelessness." *Cultural Studies* 36, no. 3 (2022): 507–9. Online ahead of print, December 2020. https://doi.org/10.1080/09502386.2020.1855455.

Pierson, George Wilson. *Tocqueville in America*. Baltimore: Johns Hopkins University Press, 1959.

Pupera, Daryl, G. "Monitoring and Enforcement of Air Quality." Department of Air Quality, Louisiana Legislative Auditor, Performance Audit Services. January 20, 2021. https://app.lla.state.la.us/PublicReports.nsf/0/4F3372ABDDF0F271862586630067C25D/$FILE/00022660A.pdf?OpenElement&.7773098.

Rabalais, Nancy, et al. "Characterization of Hypoxia: Topic I Report for the Integrated Assessment on Hypoxia in the Gulf of Mexico." NOAA Coastal Ocean Program, Decision Analysis Series 15. NOAA/National Centers for Coastal Ocean Science, Silver Spring, MD, 2000. http://oceanservice.noaa.gov/products/hypox_t1final.pdf.

Randolph, Ned. "Modeling Authority over a Drowning Coast." *Environmental Politics* 32, no. 3 (2022): 532–56. https://doi.org/10.1080/09644016.2022.2113357.

———. "River Activism, 'Levees-Only' and the Great Mississippi Flood of 1927." *Media and Communication* 6, no. 1 (February 2018): 43–51.

Ravits, Sarah. "'A Long Way to Go': Bayou Residents Still Struggle to Recover from Ida as Next Hurricane Season Looms." *Gambit*, May 13, 2022. Accessed May 17, 2023. https://www.nola.com/gambit/news/the_latest/a-long-way-to-go-bayou-residents-still-struggle-to-recover-from-ida-as-next/article_3a62ea18-d22c-11ec-8d69-ebb0a8f9a3e0.html.

Reed, Denise. Phone conversation with the author. 2018.

Reed, Denise J., and Lee Wilson. "Coast 2050: A New Approach to Restoration of Louisiana Coastal Wetlands." *Physical Geography* 25, no. 1 (2004): 4–21. https://doi.org/10.2747/0272-3646.25.1.4.

Rees, Mark, ed. *Archaeology of Louisiana*. Baton Rouge: LSU Press, 2011.

Reuss, Martin. "Andrew A. Humphreys and the Development of Hydraulic Engineering: Politics and Technology in the Army Corps of Engineers, 1850–1950." *Technology and Culture* 26, no. 1 (January 1985): 1–33. https://www.jstor.org/stable/3104527.

———. "The Art of Scientific Precision: River Research in the United States Army Corps of Engineers to 1945." *Technology and Culture* 40, no. 2 (April 1999): 292–323. https://doi.org/10.1353/tech.1999.0104.

———. "The Army Corps of Engineers and Flood-Control Politics on the Lower Mississippi." *Louisiana History: Journal of the Louisiana Historical Association* 23, no. 2 (Spring 1982): 131–48.

Ricks, George. In conversation with the author. Chalmette, LA, 2016.

———. Public comments observed by the author. Coastal Protection and Restoration Board Meeting, Houma, LA, 2019.

Rifkin, Glenn. "Overlooked No More: Homer Plessy, Who Sat on a Train and Stood Up for Civil Rights." *New York Times*, January 21, 2020. Accessed May 17, 2023. https://www.nytimes.com/2020/01/31/obituaries/homer-plessy-overlooked-black-history-month.html.

Ripple, W. J., C. Wolf, T. M. Newsome, J. W. Gregg, T. M. Lenton, I. Palomo, J. A. J. Eikelboom, B. E. Law, S. Huq, P. B. Duffy, and J. Rockström. "World Scientists' Warning of a Climate Emergency 2021." *BioScience* 71 (2021): 894–98. https://doi.org/10.1093/biosci/biab079.

Rivlin, Gary. *Katrina: After the Flood*. New York: Simon & Schuster, 2015.

———. "Why New Orleans's Black Residents Are Still Under Water after Katrina." *New York Times Magazine*, August 23, 2015. Accessed May 17, 2023. https://www.nytimes.com/2015/08/23/magazine/why-new-orleans-black-residents-are-still-under-water-after-katrina.html.

Roach, Joseph. *Cities of the Dead: Circum-Atlantic Performance*. New York: Columbia University Press, 1996.

Roberts III, Faimon. "Deadline in Plaquemines Parish–State Standoff Passes without Action." *The Advocate* (Baton Rouge), June 29, 2018. Accessed May 17, 2023. https://www.theadvocate.com/new_orleans/news/environment/article_2cd394de-7be5-11e8-88f5-73f3952a60be.html.

———. "Skull Fragment Found in the 1980s Dated to Woman Who Died 3,500 Years Ago." *New Orleans Advocate Times-Picayune*, April 30, 2023. Print.

Robertson, Campbell. "We Built an App for That. Keeping Track of the State's Lost Tombs." *New York Times*, September 26, 2016. Accessed May 17, 2023. https://www.nytimes.com/2016/09/19/us/we-built-an-app-keeping-track-of-louisianas-flood-tossed-dead.html.

Robinson, Michael. "Harold N. Fisk: A Luminescent Man." *Engineering Geology* 45, nos. 1–4 (December 1996): 37–44. https://doi.org/10.1016/S0013-7952(96)00005-1.

Rogers, J. D. "Development of the New Orleans Flood Protection System prior to Hurricane Katrina." *Journal of Geotechnical and Geoenvironmental Engineering* 134, no. 5 (May 2008). https://doi.org/10.1061/(ASCE)1090-0241(2008)134:5(602).

Rothman, Adam. "Georgetown University and the Business of Slavery." *Washington History* 29, no. 2 (Fall 2017): 18–22.

Sack, Kevin, and John Schwartz. "Left to Louisiana's Tides, a Village Fights for Time." *New York Times*, February 24, 2018. Accessed May 17, 2023. https://www.nytimes.com/interactive/2018/02/24/us/jean-lafitte-floodwaters.html.

Saikku, Mikko. "Taming the Rivers." In *Thomas Pynchon and the Dark Passages of History*, 138–64. Athens: University of Georgia Press, 2005. http://www.jstor.org/stable/j.ctt46n4v9.

Sanders, Patrick. "Blanco v. Burton: Louisiana's Struggle for Cooperative Federalism in Offshore Energy Development." *Louisiana Law Review* 69, no. 1 (Fall 2008): 255–79. http://digitalcommons.law.lsu.edu/lalrev/vol69/iss1/11.

Sandlin, Lee. *Wicked River: The Mississippi When It Last Ran Wild*. New York: Random House, 2011.

Scallan, Matt. "African American Cemeteries Plowed Over for Spillway Now Recognized as Historic." *New Orleans Advocate Times-Picayune*, January 24, 2021. Accessed May 16, 2023. https://www.nola.com/news/environment/article_cd25c180-81e1-53d8-acd1-0e5ab4973b84.html.

Schleifstein, Mark. "Dredging Mississippi River to 50 Feet Clears Corps Approval Hurdle." *New Orleans Advocate Times-Picayune*, August 20, 2018. Accessed May 17, 2023. https://www.nola.com/news/environment/dredging-mississippi-river-to-50-feet-clears-corps-approval-hurdle/article_6fe1c7fe-db4b-5a1d-866d-b4ba8b20635c.html

———. "EPA Investigates Louisiana Environmental, Health Agencies for Racial Discrimination." *New Orleans Advocate Times-Picayune*, April 7, 2022. Accessed May 17, 2023. https://www.nola.com/news/environment/epa-investigates-louisiana-environmental-health

-agencies-for-racial-discrimination-in-issuing-air-pollution-permits/article
_080b2ee6-b6a1-11ec-853a-47ef79c6ad53.html.

———. "Louisiana Granted Final Funds for Unprecedented Coastal Restoration Project." *New Orleans Advocate Times-Picayune*, March 9, 2023. Accessed May 17, 2023. https://www.nola.com/news/environment/louisiana-granted-final-funds-for-major-diversion-project/article_da97ad26-bde9-11ed-b0a0-67bce40fb68d.html.

———. "Louisiana Should Better Identify Air Pollution Violations, Speed Enforcement, Auditor Says." *New Orleans Advocate Times-Picayune*, January 20, 2021. Accessed May 16, 2023. https://www.nola.com/news/business/article_e5381f4a-5f5e-11eb-a97c-635c83e7c20d.html.

———. "Orphaned Wells Increased by 50 Percent, Could Cost State Millions: Audit." *New Orleans Advocate Times-Picayune*, April 19, 2020. Accessed May 17, 2023. https://www.nola.com/news/business/number-of-orphaned-wells-increased-by-50-percent-could-cost-state-millions-audit/article_313d8dd2-7a9d-11ea-b4a4-e7675d1484f7.html.

———. "'Power of the River': New Mississippi River Channel Prompts Corps to Take Action." *New Orleans Advocate Times-Picayune*, June 3, 2022. Accessed May 17, 2023. https://www.nola.com/news/environment/article_f56ac97c-e341-11ec-988e-77b08196f4e2.html.

Schneider, Paul. *Old Man River: The Mississippi River in North American History*. New York: Henry Holt and Co., 2013.

Scott, James. *Seeing Like a State: How Certain Schemes to Improve the Human Condition Have Failed*. New Haven, CT: Yale University Press, 1998.

Seck, Ibrahim. *Bouki Fait Gombo: A History of the Slave Population of Habitation Haydel (Whitney Plantation) Louisiana, 1750–1860*. New Orleans: University of New Orleans Press, 2014.

Shallat, Todd. "Building Waterways, 1802–1861: Science and the United States Army in Early Public Works." *Technology and Culture* 31, no. 1 (January 1990): 18–50.

———. "Holding Louisiana." *Technology and Culture* 47, no. 1 (January 2006): 102–7. https://doi.org/10.1353/tech.2006.0097.

———. *Structures in the Stream: Water, Science, and the Rise of the U.S. Army Corps of Engineers*. Austin: University of Texas Press, 1994.

Sheringham, Michael. "Archiving." In *Restless Cities*, edited by Matthew Beaumont and Gregory Dart, 1–18. New York: Verso, 2010.

Shields, Gerard. "Bill Ties La. Aid, ANWR Drilling: Relief Plan Added to Defense Budget." *The Advocate* (Baton Rouge), December 19, 2005.

Shindell, Drew (Chair). "Global Methane Assessment: Benefits and Costs of Mitigating Methane Emissions." UN Environment Program, Climate and Clean Air Coalition, May 6, 2021. Accessed May 16, 2023. https://www.unep.org/resources/report/global-methane-assessment-benefits-and-costs-mitigating-methane-emissions.

Simmonds, Julie, Sharon Lavigne, and Anne Rolfes." "Army Corps Orders Full Environmental Review of Formosa Plastics' Controversial Louisiana Plant, Decision Follows Lawsuit, Permit Suspension, Public Pressure." Center for Biological Diversity. August 18, 2021. https://biologicaldiversity.org/w/news/press-releases/army-corps-orders-full-environmental-review-of-formosa-plastics-controversial-louisiana-plant-2021-08-18/#:~:text=%E2%80%9CThe%20Army%20Corps%20has%20finally,her%20community%20from%20petrochemical%20polluters.

"Slavery in Louisiana." Whitney Plantation Museum. Accessed April 24, 2023. www
.whitneyplantation.org/history/slavery-in-louisiana.

Smethurst, Paul. "Nature Writing." In *The Routledge Research Companion to Travel Writing*,
edited by Alasdair Pettinger and Tim Youngs, 30. 1st ed. London: Routledge, 2019.

Smith, Clint. *How the Word Is Passed: A Reckoning with the History of Slavery across America*.
New York: Little, Brown, 2021.

Smith, Mike. "As Louisiana Braces for New Flood Insurance Rates, Few Understand the New
System: Steep Increases Expected for Many in the State's South, Where Water Is Never
Far Away." *New Orleans Advocate Times-Picayune*, April 3, 2022. Accessed May 13, 2023.
https://www.theadvocate.com/baton_rouge/news/article_4babe664-b1f0-11ec-9517
-1b21c3506444.html.

———. "New Data Shows Sharp Flood Insurance Hikes across South Louisiana." *New
Orleans Advocate Times-Picayune*, April 23, 2023. Print.

Smith, Mike, and Mark Schleifstein. "Corps Searches for Cause of Pump Corrosion." *New
Orleans Advocate Times-Picayune*, March 19, 2023. Print.

Smith, Sarah. "Keeping Time: Maroon Assemblages and Black Life in Crisis." *South Atlantic
Quarterly* 121, no. 1 (2022): 11–32. https://doi.org/10.1215/00382876-9561503.

Sneath, Sara. "Louisiana Hopes New Oyster Leases Will Ease Pain of Coastal Restoration
Efforts." *Times-Picayune*, June 19, 2018. Accessed May 16, 2023. https://www.nola.com
/environment/2018/06/louisiana_searches_for_ways_to.html.

———. "Louisiana Shell Refinery Left Spewing Chemicals after Hurricane Ida." WWNO
(New Orleans Public Radio), September 4, 2021. Accessed May 16, 2023. https://www
.wwno.org/science-health/2021-09-04/louisiana-shell-refinery-left-spewing-chemicals
-after-hurricane-ida.

———. "The Oil and Gas Industry Is Using Louisiana's Climate Task Force to Push Carbon
Capture." WWNO (New Orleans Public Radio), October 7, 2021. Accessed May 16, 2023.
https://www.wwno.org/news/2021-10-07/the-oil-and-gas-industry-is-using-louisianas
-climate-task-force-to-push-carbon-capture.

———. "'Ticking Time Bombs': Residents Kept in the Dark about Risks to La.'s Chemical
Plants during Storms." WWNO (New Orleans Public Radio), August 27, 2020. Accessed
May 17, 2023. https://www.wwno.org/post/ticking-time-bombs-residents-kept-dark
-about-risks-las-chemical-plants-during-storms.

Solnit, Rebecca, and Rebecca Snedeker. *Unfathomable City: A New Orleans Atlas*. Berkeley:
University of California Press, 2013.

Southeast Louisiana Flood Protection Authority–East (Flood Protection Authority).
Accessed February 10, 2018. http://ww.slfpae.com.

Southmayd, F. R. *Report of the Howard Association of New Orleans, of Receipts, Expenditures, and Their Work in the Epidemic of 1878, with Names of Contributors, etc.* New
Orleans: A. W. Hyatt, 1878. Accessed May 16, 2023. https://collections.nlm.nih.gov/catalog
/nlm:nlmuid-9711174-bk.

Spears, Nancy Marie. "Freedmen Descendants Hopeful for Increased Federal Care." *Gay-
lord News*, October 29, 2021. Accessed May 16, 2023. https://ictnews.org/news/freedmen
-descendants-hopeful-for-increased-federal-care.

Staff Editorial. "Our Views. A New Wave of Industry Growth." *The Advocate* (Baton Rouge),
December 10, 2018. Online ahead of print, accessed November 28, 2018. https://www

.theadvocate.com/baton_rouge/opinion/our_views/article_2d9ec46c-f401-11e8-a5ec
-cb526d534750.html.

———. "Our Views: Upcoming Petrochemical Projects in Louisiana Mean More Construction Jobs." *The Advocate* (Baton Rouge), March 3, 2019. Accessed May 4, 2022. https://
www.theadvocate.com/baton_rouge/opinion/our_views/article_e179d4c8-c90c-11e9
-acf6-ab095b43cb8b.html.

———. "Our Views: Welcome New Investment in Louisiana's Petrochemical Industry."
The Advocate (Baton Rouge), December 22, 2020. Online ahead of print, accessed
December 15, 2020. https://www.theadvocate.com/baton_rouge/opinion/our_views
/article_4e6c730c-3fe9-11eb-ac23-3f45dade76cc.html.

State Library of Louisiana Historic Photograph Collection. "Sugar Kettle outside of the
Chemical Engineering Building at Louisiana State University in Baton Rouge Louisiana
circa 1969." Accessed May 4, 2022. https://louisianadigitallibrary.org/islandora/object
/state-lhp%3A1058.

State of Louisiana Executive Order Number JBE2020-18, Climate Initiatives Task Force.
Accessed May 4, 2022. https://gov.louisiana.gov/assets/ExecutiveOrders/2020/JBE-2020
-18-Climate-Initiatives-Task-Force.pdf.

Steiger, Brad. *The Werewolf Book: The Encyclopedia of Shape-Shifting Beings.* Canton, MI:
Visible Ink Press, 2012.

Striphas, Ted. "Caring for Cultural Studies." *Cultural Studies* 33, no. 1 (2019): 1–18. https://
doi.org/10.1080/09502386.2018.1543716.

"Subsidence." City of Long Beach. Accessed September 27, 2018, http://www.longbeach.gov
/lbgo/about-us/oil/subsidence.

"Sugar at LSU: A Chronology." Exhibition curated by Christina Riquelmy (Special Collections Rare Book Cataloger) and Debra Currie (Reference Librarian, Middleton Library).
Accessed May 4, 2022. https://www.lib.lsu.edu/sites/all/files/sc/exhibits/e-exhibits/sugar
/contents.html.

Swanton, John R. "Hernando de Soto's Route through Arkansas." *American Antiquity* 18,
no. 2 (October 1952): 156–62. https://doi.org/10.2307/276540.

Swyngedouw, Erik, Maria Kaka, and Esteban Castro. "Urban Water: A Political-Ecology
Perspective." In "Water Management in Urban Areas," special issue, *Built Environment*
28, no. 2 (2002): 124–37.

Szeman, Imre, and D. Boyer. Introduction to *Energy Humanities: An Anthology*, edited by
Imre Szeman and D. Boyer. Baltimore: Johns Hopkins University Press, 2017.

Tahir, Sheila. In conversation with the author. Down by the River Bike Ride, Norco, LA,
April 2022.

Taylor, Dorceta E. *Toxic Communities: Environmental Racism, Industrial Pollution, and
Residential Mobility.* New York: New York University Press, 2014.

Terrell, Kimberly A., and Wesley James. "Racial Disparities in Air Pollution Burden and
COVID-19 Deaths in Louisiana, USA, in the Context of Long-Term Changes in Fine
Particulate Pollution." *Environmental Justice* 15, no. 5 (2022): 286–97. Online ahead of
print, 2020. https://doi.org/10.1089/env.2020.0021.

Terrell, Kimberly A., and Gianna St Julien. "Air Pollution Is Linked to Higher Cancer Rates
among Black or Impoverished Communities in Louisiana." *Environmental Research Letters* 17, no. 1 (2022). https://doi.org/10.1088/1748-9326/ac4360.

Tessum, C. W., J. S. Apte, A. L. Goodkind, N. Z. Muller, K. A. Mullins, D. A. Paolella, et al. "Inequity in Consumption of Goods and Services Adds to Racial-Ethnic Disparities in Air Pollution Exposure." *Proceedings of the National Academy of Sciences* 116, no. 13 (2019): 6001–6. https://doi.org/10.1073/PNAS.1818859116.

Theriot, Jason. *American Energy, Imperiled Coast: Oil and Gas Development in Louisiana's Wetlands*. Baton Rouge: LSU Press, 2014.

Thompson, Richard. "Louisiana Ports Support More Than 396,000 Jobs in the State, Study Shows." *New Orleans Advocate Times-Picayune*, May 22, 2021. Accessed May 17, 2023. https://www.nola.com/news/business/louisiana-ports-support-more-than-396-000-jobs-in-the-state-study-shows/article_016eca17-ae30-5bd5-bbe2-13793e46a486.html.

Thompson, Shirley. "No Sweetness Is Light." In *Unfathomable City: A New Orleans Atlas*, by Rebecca Solnit and Rebecca Snedeker, 72. Berkeley: University of California Press, 2013.

Thorington, Brooke. "Mid-Barataria Sediment Diversion Project Could Devastate the Seafood Industry in Louisiana." Louisiana Radio Network, January 17, 2023. Accessed May 17, 2023. https://louisianaradionetwork.com/2023/01/17/chair-of-louisiana-oyster-task-force-says-sediment-diversion-project-will-wipe-out-seafood-industry-in-louisiana/.

Thunberg, Greta. "'How Dare You' Speech to the United Nations." YouTube, 2019. Accessed May 16, 2023. https://youtu.be/xVlRompc1yE.

Together Louisiana. "The Biggest Corporate Welfare Program in the Nation." Accessed April 26, 2022. https://www.togetherla.org/fairtaxes.

Tripp, James, and Michael Herz. "Wetland Preservation and Restoration: Changing Federal Priorities." *Virginia Journal of Natural Resources Law* 7, no. 2 (Spring 1988): 221–75.

"Trump: 'Why would I care about the climate? I'll be dead in 10 years anyway.'" *The Postillion*, June 2, 2017. Accessed May 17, 2023. https://www.the-postillon.com/2017/06/trump-interview-paris-climate.html.

Tsing, Anna. *The Mushroom at the End of the World: On the Possibility of Life in Capitalist Ruins*. Princeton, NJ: Princeton University Press, 2021.

Turner, R. Eugene. "Discussion of: Olea, R. A. and Coleman, J. L., Jr., 2014 . . . " *Journal of Coastal Research* 30, no. 6 (November 2014): 1330–34. https://www.jcronline.org/doi/pdf/10.2112/JCOASTRES-D-14-00076.1.

———. "Doubt and the Values of an Ignorance-Based World View for Restoration: Coastal Louisiana Wetlands." *Estuaries and Coasts* 32, no. 6 (November 2009): 1054–68. https://doi.org/10.1007/s12237-009-9214-4.

———. In conversation with the author. 2017.

———. "The Mineral Sediment Loading of the Modern Mississippi River Delta: What Is the Restoration Baseline?" *Journal of Coastal Conservation* 21, no. 6 (2017): 867–72. https://doi.org/10.1007/s11852-017-0547-z.

Turner, R. Eugene, and Giovanni McClenachan. "Reversing Wetland Death from 35,000 Cuts: Opportunities to Restore Louisiana's Dredged Canals." *PLoS ONE* 13, no. 12 (December 2018): e0207717. https://doi.org/10.1371/journal.pone.0207717.

Twain, Mark. *Life on the Mississippi*. Toronto: Musson, 1883; Project Gutenberg, 2006, last updated 2018.

US Army Corps of Engineers. "Notice of Intent to Prepare a Draft Environmental Impact Statement for the Lake Pontchartrain and Vicinity General Re-evaluation Report,

Louisiana." *Federal Register* 84, no. 63 (April 2, 2019): 12598–99. https://www.govinfo
.gov/content/pkg/FR-2019-04-02/pdf/2019-06354.pdf.

———. "Andrew Atkinson Humphreys (1810–1883)." Accessed June 14, 2019. https://www
.mvn.usace.army.mil/About/History/Army-Engineers-in-the-Civil-War/Engineer
-Biographies/Andrew-Humphreys/.

———. "Welcome to the J. Bennett Johnson Waterway." Accessed June 3, 2022. https://www
.mvk.usace.army.mil/Missions/Recreation/J-Bennett-Johnston-Waterway/.

———. New Orleans District. "Old River Control." Accessed June 14, 2019. https://www
.mvn.usace.army.mil/Missions/Recreation/Old-River-Control/.

US Energy Information Administration. "Louisiana State Energy Profile." May 19, 2022.
Accessed May 16, 2023. https://www.eia.gov/state/print.php?sid=LA.

US Environmental Protection Agency (EPA). "National Air Toxics Assessment; Request
for Reconsideration, Denka Performance Elastomer LLC." Washington DC, 2018.
Accessed May 12, 2023. https://www.epa.gov/sites/default/files/2018-08/documents
/rfr_exhibits_a-g_n3630829x7a3a0.pdf.

———. "What Climate Change Means for Louisiana and the Gulf Coast." Accessed April 24,
2023. https://www.epa.gov/sites/production/files/2016-09/documents/climate-change
-la.pdf.

US Government Accountability Office (GAO). "Offshore Oil and Gas: Updated Regulations
Needed to Improve Pipeline Oversight and Decommissioning." GAO-21-293. March 19,
2021. Accessed May 17, 2023. https://www.gao.gov/products/gao-21-293.

Vaughan, Elizabeth. "Louisiana Sugar: A Geo Historical Perspective." PhD diss., Louisiana
State University and Agricultural and Mechanical College, 2003. Accessed May 17, 2023.
https://digitalcommons.lsu.edu/cgi/viewcontent.cgi?article=4692&context=gradschool
_dissertations.

Verchick, Rob, Monique Harden, and Karen Sokol. "The False Promise of Carbon Capture
in Louisiana." Center for Progressive Reform and DeepSouth Center for Environmental
Justice, March 10, 2020. Accessed May 16, 2023. https://cpr-assets.s3.amazonaws.com
/documents/false-promise-ccs-louisiana-webinar-slides-031022.pdf.

The Water Campus. Accessed July 26, 2018. http://www.thewatercampus.org.

Watts, Sheldon. "Yellow Fever, Malaria and Development: Atlantic Africa and the New
World, 1647 to 1928." In *Epidemics and History: Disease, Power and Imperialism*, 213–68.
New Haven, CT: Yale University Press, 1997. http://www.jstor.org/stable/j.cttnq8qw.11.

Whayne, Jeannie, Thomas A. DeBlack, George Sabo III, Morris S. Arnold, and Orrin H.
Ingram. "Spanish and French Explorations in the Mississippi Valley." In *Arkansas: A
Narrative History*, 39–52. Fayetteville: University of Arkansas Press, 2013.

Wicker, Karen, et al. "Mississippi Deltaic Plain Region Ecological Characterization: A Habitat Mapping Study: A User's Guide to the Habitat Maps." FWS/OBS-79/07. US Fish and
Wildlife Service, Office of Biological Services, May 1980. Accessed May 17, 2023. https://
catalog.hathitrust.org/Record/007402320.

Wilkins, Jim, et al. "Preliminary Options for Establishing Recreational Servitudes for
Aquatic Access over Private Water Bottoms." Submitted by the Louisiana Sea Grant
College Program in response to House Resolution 178 of the Louisiana Legislature.
Louisiana Sea Grant, March 1, 2018. Accessed May 17, 2023. http://www.laseagrant.org
/wp-content/uploads/LSG-Coastal-Access-Report.pdf.

Williams, B. "'That we may live': Pesticides, Plantations, and Environmental Racism in the United States South." *Environment and Planning E: Nature and Space* 1, nos. 1–2 (2018): 243–67. https://doi.org/10.1177/2514848618778085.

Williams, Jeffress. "Louisiana Coastal Wetlands: A Resource at Risk." U.S. Geological Survey Fact Sheet. Accessed September 28, 2018. https://pubs.usgs.gov/fs/la-wetlands/.

Williams, Thomas Chatterton. *Self-Portrait in Black and White: Family, Fatherhood and Rethinking Race.* New York: W. W. Norton, 2019.

"William Tecumseh Sherman." LSU Military Museum, Baton Rouge. Way Back Machine. Accessed May 17, 2023. https://web.archive.org/web/20221009004725/http://olewarskule .lsu.edu/lsu-military-museum/william-tecumseh-sherman.html.

Wright, J. O. *Swamp and Overflowed Lands in the United States: Ownership and Reclamation.* US Department of Agriculture, Office of Experiment Stations, Circular 76. Washington, DC: Government Printing Office, 1907. https://doi.org/10.5962/bhl.title.87650.

Yeomans, Jon, and Donna Bowater. "One Year On, Brazil Battles to Rebuild after the Samarco Mining Disaster." *Telegraph*, October 15, 2016. Accessed May 17, 2023. https:// www.telegraph.co.uk/business/2016/10/15/one-year-on-brazil-battles-to-rebuild-after -the-samarco-mining-d/.

Yuill, Brendan, Dawn Lavoie, and Denise J. Reed. "Understanding Subsidence Processes in Coastal Louisiana." *Journal of Coastal Research*, no. 54 (2009): 23–36. https://www .jcronline.org/doi/full/10.2112/SI54-012.1.

INDEX

Abbot, Henry L., 84
abolition movement, 158, 163
Act 8 (Louisiana), US (2005), 129
Act 41 (Louisiana), US (1981), 112
African Americans: diseases and infections
    for, 55; environmental redlining of, 164–65;
    Freedmen and, 51; freetowns for, 164–65,
    172; in LULUs, 165–68; Native Americans
    and, 51–52. *See also* institutionalized racism;
    racial capitalism; racial classification and
    designation; racial identity
African Queen, The (Forster), 43–44
AGA. *See* American Gas Association
Alabama: enslaved persons in, 80; white
    population in, 80
Alexandria, Louisiana, during US Civil War, 1
alluvial rivers, 13, 76; morphology of, 94–95
American Gas Association (AGA), 126–27
American Journal of Science, 84
American Red Cross, 89
American Revolutionary War, Swamp Foxes
    during, 44
America's WETLANDS campaign, 120–21,
    125–26, 137
Anderson, Charles, 94–95
Anilco settlement, 23
Anthropocene: capitalism and, 11; climate crisis,
    183; definition of, 10; environmental equity
    strategies and, 184; Eurocentrism and, 184;
    globalization and, 11; Great Acceleration of

consumption during, 10; whiteness
    and, 184
ANWR. *See* Arctic National Wildlife Refuge
Arctic National Wildlife Refuge (ANWR),
    126–27
Arkansas River, 22–23

backfilling, of canals: oil and gas production
    and, 103–7; opposition to, 137; partial,
    105; purpose of, 106, 114; with spoil
    banks, 106
Bahr, Len, 105–6
Bailey, Joseph, 1
"Bailey's Dam," 1
Baker, Richard, 128, 147
Barataria Diversion, 140–42
Barataria-Terrebonne National Estuary Program
    (BTNEP), 115
Barry, John, 85, 88
Baton Rouge, Louisiana: oil and petroleum
    production corridor, 14–15. *See also*
    Cancer Alley
Battle of New Orleans, 44
Bentham, Jeremy, 53
Bernoudi, François Bernard, 161
Biden, Joe, 151, 177–78
Bienville, Jean-Baptist Le Moyne, Sieur de, 34
Big Muddy. *See* Mississippi River
Bird's Foot Delta, 30, 35–36
Black, William M., 21

Founded in 1893,
UNIVERSITY OF CALIFORNIA PRESS
publishes bold, progressive books and journals
on topics in the arts, humanities, social sciences,
and natural sciences—with a focus on social
justice issues—that inspire thought and action
among readers worldwide.

The UC PRESS FOUNDATION
raises funds to uphold the press's vital role
as an independent, nonprofit publisher, and
receives philanthropic support from a wide
range of individuals and institutions—and from
committed readers like you. To learn more, visit
ucpress.edu/supportus.